CLOSING THE LOOP

The Story of Feedback

Stanley W. Angrist

Closing the Loop

The Story of Feedback

DRAWINGS

BY ENRICO ARNO

THOMAS Y. CROWELL COMPANY / NEW YORK

Manufactured in the United States of America

ISBN 0–690–19644–X

1 2 3 4 5 6 7 8 9 10

Library of Congress Cataloging in Publication Data
Angrist, Stanley W.
  Closing the loop.

  Bibliography; p.
  1. Feedback control systems. I. Title.
TJ216.A54 1973   629.8'3   73-3
ISBN 0-690-19644-X

To my mother

Contents

CLOSING THE LOOP

The Story of Feedback

1 What Is Feedback?

EXCEPT for the dim glow of the instrument lights the flight deck onboard Pan American Airways Flight Number 63 from London to New York is dark. The giant Boeing 747 is at 30,000 feet and moving along at a brisk 625 miles per hour. The passengers, all three hundred twenty of them, expect that in less than thirty minutes they will be arriving at John F. Kennedy International Airport to be greeted by relatives and friends who are waiting for them.

Before he left London, however, the captain of Flight Number 63 had been informed that the weather in New York, at his scheduled arrival time, might be touch and go. A big snowstorm was moving in from the midwest and it could cause JFK, as the airport is called, to close down for the night. An airport closes to incoming traffic when the runway visual range (called the RVR) drops below a certain minimum distance. This distance is measured by an electric eye located at the end of the runways used by airplanes that are landing. When the electric eye cannot "see" a light located 1,600 feet away, JFK must not allow planes to land until the weather improves enough to give an RVR of at least 1,600 feet.

A message begins to crackle through the captain's headphones: "This is New York Air Route Traffic Control (ARTC) to Pan Am 63. Please be advised that due to heavy snow the RVR at Kennedy is now below the minimum set by the Federal Aviation Agency and thus JFK is now closed to all arriving flights; it will probably remain closed for at least three hours and perhaps longer. What are your intentions?"

Before the captain of Pan Am Flight Number 63 left London, he listed on his flight plan an alternate airport at which he might land in the event that JFK closed. The captain informs the New York ARTC Center that his alternate airport is Boston and asks for a course and altitude that will take him there. The New York Center obliges and then telephones the Boston Air Route Traffic Control Center; the Boston Center will now control Pan Am 63 on the final portion of its flight. The captain swings the big plane on a heading toward Boston and tells his anxious passengers their new destination. Several hundred passengers and their waiting friends and relatives will be greatly disappointed upon hearing of Pan Am 63's new destination.

Why did JFK have to close down to incoming planes this night? Is there any hope that man will be able to land his airplanes safely in even the worst weather? The answer to both of these questions, surprisingly enough, lies in one word—*feedback*.

Feedback has been defined as the property of being able to adjust future conduct by past performance. It is exactly this property that is required to land an airplane—whether the landing be done by a pilot or an automatic landing system. Man is now learning to build into his machines the ability both to seek goals and to correct themselves while seeking those goals. For thousands of years it was believed that only living organisms could exhibit these two characteristics.

But what role does feedback play in the landing of Pan Am Flight Number 63? Furthermore, what role can it play in the future use of automatic equipment? We can find out by thinking about what exactly a pilot does when he lands an airplane.

Before and during the landing of an airplane the pilot receives considerable information about his plane and about conditions on the ground. By radio from the control tower he knows the wind speed and direction on the ground; from instruments onboard the plane he knows his own speed and direction, his own attitude (whether the nose is above the horizon, even with it, or below it), how fast he is losing altitude, and so on. He can, under conditions of good weather, see the runway where he is going to land. Based on how the runway appears to him and what his instruments tell him, he aligns the airplane in three ways as he approaches. If he is

FIGURE I

At night under good weather conditions a pilot of a Boeing 747 has this view of the runway before he actually begins his final descent. During final approach the long horizontal bar which is located on a small plastic plate fixed in front of the windshield (shown in the middle of the photograph) would be aligned with the short, bright, horizontal bar which is the "threshold" or beginning of the runway on which the plane will be landed. Used properly this approach monitor guides the pilot to the appropriate final approach angle for a good landing. Photograph courtesy of Pan American World Airways.

landing too fast, he cuts back the thrust of the engines; if too slow, he increases the thrust of the engines (by increasing the fuel flow to the engines); if he is coming in too steeply, he raises the nose; too shallow, he lowers the nose. While all this is going on he keeps the plane aligned with the runway so that it is neither too far to the left nor too far to the right. Finally, by using all the information presented by the airplane's instruments, along with his own eyes and years of experience, he places the 500,000 pounds of airplane on the runway at a speed of 160 miles per hour. During a good landing the passengers may not even be certain when the wheels actually touch the runway.

But the captain of Pan Am Flight Number 63 was not permitted to complete these procedures and land his plane at JFK on this night. Why? The answer to this question is found in the Federal Aviation Agency's regulation book which states under what weather conditions an airport must close. The FAA believes that when a pilot's range of vision is less than 1,600 feet at the end of the runway, he simply might not have enough time to make the corrections that are required in adjusting the airplane's altitude, heading, speed, or attitude to achieve a good landing. Man, as he completes the necessary procedures required to land an airplane, might simply be too slow or too inaccurate.

Now it is possible to build an electronic-mechanical system that will do everything that the pilot must do and more. That is, an automatic landing system can take over the control of the airplane when it is twenty miles or so away from the airport, align the plane with the runway while reducing its altitude, and finally, bring the wheels of the plane down onto the runway at just the right speed. All of this can be done while the pilot sits with his hands folded in his lap—hence the informal name for such a system is a "hands off" landing system. The automatic landing system works by aligning itself with radio signals sent out from transmitters located near the runway. These signals are then fed into instruments which operate the rudder, elevators, ailerons, and engine throttles on the airplane. These instruments essentially replace the pilot's hands on the controls.

FIGURE 2

A Boeing 747 cockpit showing a pilot's (at the left) and co-pilot's instrument panels. The round dials in the upper center of the photograph are engine instruments which monitor temperature, speed, fuel flow, and the like. Instruments showing the aircraft's heading, attitude and position are directly in front of the pilot and co-pilot. Engine throttles are located directly between the pilot and co-pilot (numbered 1, 2, 3, 4 in the photograph). Communications equipment is shown in the foreground. Photograph courtesy of Pan American World Airways.

Instead of correcting mistakes in the airplane's heading or altitude based on what is seen through the cockpit windows, the landing system judges how the airplane is doing with respect to the radio signals it receives. It is only because the system is built around a system which is able to check and correct itself that a feedback machine is able to land a 500,000 pound airplane at 160 miles per hour.

Today, during the landing process, a pilot exhibits goal-seeking behavior by using his body's feedback systems, which have been developed to a very high level of skill through years of training and practice. For it is his job to collect information—mostly with his eyes—and then use his accumulated years of experience and training to send appropriate signals to his hands and feet which will, by way of the airplane's controls, bring the wheels of the plane into contact with the runway. In this man-machine feedback system it is the pilot who does the checking and correcting that causes the whole system to achieve the desired goal.

If you wanted to build a machine that would roll a bowling ball down the gutter that runs along the side of a bowling alley, no feedback system or loop would be required. Once the initial force needed to get the ball to the end of the gutter had been determined

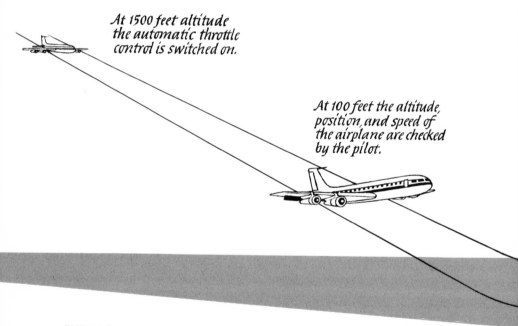

*At 1500 feet altitude
the automatic throttle
control is switched on.*

*At 100 feet the altitude,
position, and speed of
the airplane are checked
by the pilot.*

FIGURE 3
A schematic illustration of what a completely automatic ("hands-off") landing system must do during the landing process.

and built into the machine, it would roll balls down the gutter flaw-
lessly all day. One of the reasons no feedback system is required
is that the bowling ball is not likely to be surprised by any unexpected occurrences; in landing a huge airplane unexpected things
have a way of happening very frequently—a gust of wind shoves
the plane away from its alignment with the runway just as it gets
over the edge, or the wind speed drops and the plane suddenly
noses down in response to nature. Any landing system not built
to respond to the unexpected would do nothing but produce
cracked-up airplanes.

An automatic landing system, like any pilot, is said to exhibit
goal-seeking behavior, the goal being the landing of an airplane
under a wide variety of conditions. It is able to achieve its goal
because such a system can measure or judge how well it is doing
(where the airplane is, for example) with how well it should be
doing (where the airplane should be). The difference between
these bits of information is called the *error;* and it is the error that
is fed back into the system so that it might correct itself. For
example, if the plane should be on a heading of 270 degrees (due
west) in order to be aligned with the runway but is actually heading 260 degrees (slightly south of due west), a control signal will

At 50 feet the throttles are automatically closed and final landing attitude is set.

At touchdown, the automatic equipment is disconnected. Pilot uses reverse thrust to slow plane and takes over steering and braking.

be sent to a motor that operates the rudder to steer the plane onto the correct heading. At the same time, the landing system is also correcting the altitude and rate of descent needed to achieve a perfect landing.

But isn't that exactly what a pilot does? Of course, but man has now learned to build machines that he can no longer trust to man —a statement which might sound like a strange contradiction— but true nonetheless. It is simply impossible for a pilot to react to surprises fast enough when he tries to bring a 500,000-pound collection of aluminum and electronics traveling at 160 miles per hour gently into contact with a concrete runway when he can't be absolutely certain where that runway is. Though man is an excellent feedback mechanism there are now some jobs which can be done much better by machines—sometimes because man is too slow, sometimes because he is not accurate enough, and sometimes because the job must be done over and over again and he would soon become bored and careless in his work.

It is now clear that machines can be given goal-seeking properties and the ability to correct themselves while seeking those goals. And while man is inventing ways to duplicate this goal-seeking behavior in his devices, he is also finding out that by studying feedback principles he is learning more about some of the control systems in his body and he is also finding out more about how animals, including man, learn. In brief, then, his increased knowledge of feedback is being fed back so that he is understanding more about how his body functions and learns.

Besides being present in every living organism as well as in countless machines, feedback also appears in nearly every social system that man has created. As you shall learn it is feedback that keeps your body temperature at 98.6°F, your house at 68°F, that maintains a balance between predatory animals and their prey; it is also feedback that prevents stock prices from soaring off into the heavens in good times or dropping to zero in bad times. Feedback is truly one of the few universal activities found throughout nature.

2 Inputs, Outputs, Comparers, and Reducers

SUPPOSE you decide that you want to learn to shoot baskets with a basketball. If you think about that job for a moment you might decide that it is not too much different than when a pilot learns to land an airplane. You gather information with your eyes about where the basket is, then your brain orders the muscles in your arms to throw the ball. After the shot your eyes compare where the ball went to where you wanted the ball to go—which is through the basket, of course. If you missed the basket, your brain will send new instructions to your arms when you attempt the second shot, which will probably be closer than the first. After a while you will start to make baskets. Your body's feedback system was able to utilize the *error*—that is, the difference between where the basket is and where the ball actually strikes the basket or backboard—to improve your performance on the next shot. In fact, nearly all forms of learning depend upon feedback loops within the body. Learning, as it depends on feedback, will be discussed in greater detail in Chapter 6.

Systems which depend upon feedback to carry out their goals are frequently called *closed-loop* because how the system is actually doing (called the *output*) depends upon what is actually desired (which is called the *input*). A feedback

loop is formed when a portion of the output is fed back and compared with the input. Systems in which the output in no way depends upon the input, such as the one described earlier of bowling
balls rolled down the gutter, are called *open-loop* systems. In
analyzing a system to determine whether it is closed- or open-loop,
it is helpful to make simple sketches of the system as will be done
in the following pages.

Even before you were a year old, you learned that if you pushed
a ball with your hand it would roll across the floor. Your push
was recognized by you to be the "cause" of the ball's motion, and
the motion itself was the "effect" of your push. The physical world
is full of events which are connected like this, though many of
them are not nearly so obvious as this example. Scientists and
engineers are constantly trying to determine the exact nature of
the relationship between apparent cause-and-effect events. Indeed,
it is impossible to understand even the most elementary ideas surrounding feedback without understanding something about cause
and effect.

Suppose the temperature in your house is controlled by a thermostat that regulates the flow of fuel that is burned to produce
heat in a furnace. Next to the thermostat is an ordinary mercury-
in-glass thermometer. As the house becomes warmer, the thermometer indicates a higher temperature. Even a person who does not
know anything about furnaces, thermostats, and feedback would
not expect to make the house warmer by simply holding a lighted
match under the bulb of mercury on the thermometer. That is,
fooling the thermometer by raising its temperature with a match
will not warm in the least a person sitting across the room from
the thermometer, no matter what the lighted match makes the thermometer read. But now consider what a thermostat does in com-
parison with the ordinary thermometer. Thermostats are constructed
in such a way that if the house temperature drops below some
desired level, the fuel supply to the furnace is electrically switched
on, thus causing the furnace to supply heat to the house. Once the
desired temperature is reached, the thermostat senses it and switches

off the fuel supply to the furnace. This arrangement constitutes a closed-loop, or feedback system made up of the house, thermostat, and furnace. The thermometer itself is not a part of the system as it merely indicates how well the system is functioning. The temperature indicated by the thermometer does depend, of course, on how warm the room is. If the room temperature is below the desired temperature set on the thermostat, the furnace will be asked to supply heat to the room to raise its temperature to the desired level. The two quantities—the reading on the thermometer and the warmth of the room—are related in that each of them is a cause and each an effect of the other. That is characteristic of all closed-loop feedback systems.

It is now fairly easy to identify by name those things which must be present in a feedback system if it is to work successfully. At least four items are present in any such system: an *input,* goal, or desired quantity, which in the previous example is the temperature one wishes to maintain in the room; an *output* or result, which in the example is the actual temperature of the house; a *comparer* which is designed to see how close the output matches the input, and, if they are different, do something about that difference by causing a *reducer* to lessen the difference between the desired input and the actual output. (In general feedback terminology, the comparer and reducer are sometimes considered to be a single unit, called a *controller.*) In the case of the heating system described above, the comparer is the thermostat in the house which compares by a simple mechanism the actual temperature in the house and the desired temperature. The reducer is the furnace that the thermostat can switch on as the house temperature falls below the desired temperature, and switch off when the house temperature rises above the desired temperature.

Figure 4 is a sketch of the heating system showing how the actual temperature is fed back into the thermostat, where it is then compared with the desired temperature. Engineers and scientists who work with feedback systems generally don't try to draw the details of what is inside a comparer or a reducer, but simply show such

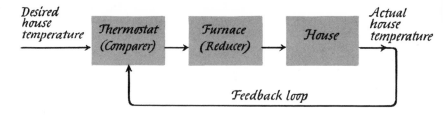

FIGURE 4
The feedback control loop for a house using a thermostatically controlled furnace. The actual temperature is the fed-back quantity telling the thermostat whether the house needs additional heat to achieve the desired temperature.

devices as boxes or blocks. Inputs and outputs are almost always represented as lines with arrowheads on them.

Frequently the term "negative feedback" is used to describe goal-seeking systems. The word negative means that the system is designed so that the comparer always subtracts the output from the desired input or takes the difference between them and then acts to reduce this difference. Whenever the word feedback is used in this book, it will be shorthand for the phrase "negative feedback" as that is the principle which has the most interesting properties and exhibits itself throughout nature.

Many feedback systems have more than one input, output, comparer, and reducer. Indeed more than one is probably the rule rather than the exception. The system used to control body temperature, for example, has as its primary output internal body temperature, but also has a secondary output of skin temperature. Each of these outputs requires its own feedback loop, comparer, and reducer. However, until you are used to thinking about feedback loops, it will be useful to concentrate on rather simple systems that contain only one input, output, comparer, and reducer.

A mechanism designed inadvertently to have "positive feedback," or that because of accident becomes a positive feedback system, can produce striking and, sometimes, catastrophic results. In these cases the comparer and reducer would act in such a way to *increase* the difference between the input and the output. For example, positive feedback sometimes occurs in public address systems; it makes its presence known by means of a high-pitched whistling sound which is generated when the microphone picks up some of the sound coming from the loudspeakers and amplifies it. The newly amplified sound is then received by the microphone/amplifier system and amplifies it again. The sound quickly builds up to a high-pitched whistle that can be quite painful. Positive feedback systems clearly do not illustrate goal-seeking behavior.

It is also possible to design control systems that have no feedback at all—neither negative nor positive. Such arrangements are called open-loop systems. Suppose this time that a heating system is designed with the thermostat on the outside of the house and not on the inside as was the case before. One might construct such an arrangement because it is reasoned that as the outside temperature

FIGURE 5
There is no feedback in this open-loop system. The thermostat located outside the house has no way of knowing if it is sending appropriate signals to the furnace. The thermostat could be calling for heat based on the outside temperature, even though the house is at a quite comfortable temperature.

starts to fall, the house will need more heat to keep the inside temperature at a reasonable level. But the thermostat in this case never compares the inside temperature with what is desired. Such systems can work if they are very carefully adjusted; but they respond poorly to unexpected disturbances or factors which are difficult to build into an adjustment. For example, heating a house with a lot of windows on its south side could be troublesome with such a system. In the winter on a sunny day, such a house might need relatively little heat as the sunshine pouring through the windows would have a tendency to maintain the inside temperature. However, on an overcast day such a house would probably need considerable heat. A thermostat located inside the house has no trouble handling either of these situations by simply comparing the desired and actual temperature in the house. But in an open-loop system, with the thermostat located outside the house, there is no way of ensuring a comfortable temperature inside. Now suppose a cold front starts to move through the area and the sun is still shining. The outside temperature begins to fall so the thermostat calls for heat; however, with the sun shining through the windows there is no need for heat probably until near sundown. The house will thus become uncomfortably warm because it is now receiving both furnace heat and sunshine. In such systems the thermostat does not have supplied information about the actual quantity (the inside temperature) that it is trying to control. Thus it can be seen that open-loop systems can be made very simply but in many cases will suffer from an inability to achieve a desired goal except by accident. (That is not to say that all open-loop systems are failures. Most home washing machines are open-loop in design in their inability to determine how clean the clothes are at the end of a washing cycle, and yet their performance in most cases is very satisfactory.)

That you yourself represent a very complicated negative feedback control system has been recognized by scientists for a number of years. Such a system can be presented in a block diagram without great difficulty. Suppose that you decide to point your finger at an airplane flying across your field of view. It is easy to identify the four elements discussed previously. The input is the precise direction you must point your finger if it is to be aligned with the airplane. The output is the direction you are actually pointing your finger. (Remember that the difference between these two directions is the error.) The comparer in this case can be considered to be your eyes and the part of your brain that receives and interprets information from them. The reducer is that part of your brain which sends signals to the muscles in your arm, hand, and finger, and those muscles themselves. It is possible that you might be using some information from your ears which could also be considered to be

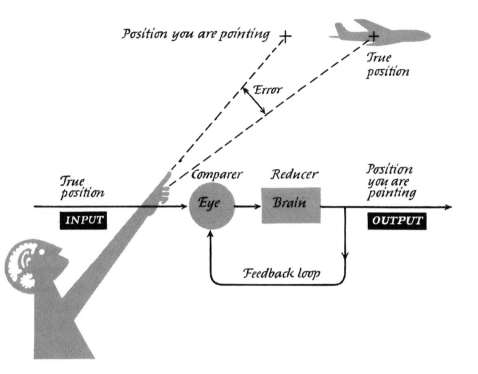

FIGURE 6
When you use your finger to point at a moving airplane, you are forming
a rather complicated feedback loop. Your eyes (comparer) and brain (reducer) enable you to keep your finger pointed reasonably close (output) to
the true location (input) of the airplane.

part of your comparer. In fact, if it is a slow airplane you could
probably do pretty well just by using your ears with your eyes
closed; however, if it is an airplane flying near the speed of sound
your ears will actually mislead you because the sound reaches you
long after the airplane has passed the spot where the sound orig-
inated.

The presence of negative feedback in a system can thus be estab-
lished if you can identify an input, an output, a comparer, and a
reducer and determine that the output is always being fed back into
the comparer in order to cause the difference between the output
and input to be reduced.

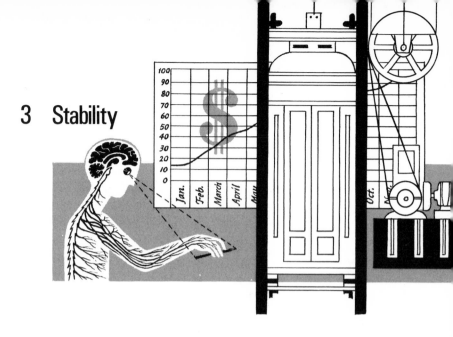

3 Stability

THE IDEA of stability is firmly fixed in our everyday language. A person who is not cool, has a hot temper, and who acts without thinking of the consequences of his actions is said to be unstable. His behavior is erratic and unpredictable. The person who carefully considers the consequences of any actions he might take, who acts cautiously and with great deliberation is said to be as "stable as the Rock of Gibraltar." It is necessary when considering the design of machines and the actions of living organisms to go beyond popular usage and look a little more deeply into the concept of stability.

Kinds of Equilibrium

The words *equilibrium* and *stability* are sometimes used interchangeably though, in fact, they do not mean the same thing. An object or system which is in equilibrium has no unbalanced forces acting upon it. An object's stability refers to how it responds when a force is applied to it. Consider, for example, a ball sitting on a flat table top as shown in Figure 7(a). A push applied to the ball will cause it to move either a lot or a little depending on the size of the push. Such a state of affairs is called *neutral equilibrium*.

Now consider the case where the ball is balanced precariously on top of a bigger ball as shown in Figure 7(b). Even a slight push will send the ball rolling off with considerable speed. This situation is described as *unstable equilibrium*. While it is true that the ball has no unbalanced forces on it, a little push quickly and permanently removes it from a state of balance.

The last situation to be considered is when the ball is resting at the bottom of a bowl as shown in Figure 7(c). A slight push disturbs the ball's state of balance but the sloping sides of the bowl quickly bring the ball back into balance when the disturbing force is removed. The ball is said to be in *stable equilibrium*. In the following pages it will be shown that feedback dependent systems in unstable and stable equilibrium behave differently when a disturbance is applied to them.

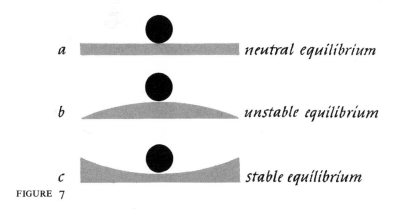

a   neutral equilibrium

b   unstable equilibrium

c   stable equilibrium

FIGURE 7

Skyscrapers Beget Skyscrapers

An expression frequently heard in cities around the world is that "skyscrapers beget skyscrapers." What does this curious expression mean? How does it relate to the notion of feedback?

Until about the turn of the century, office buildings were generally no more than five or six stories in height. There were good reasons for this limit. First, up until that time such buildings were mostly constructed of stone with the walls at each floor generally having to carry the weight of all the floors above that floor. Thus the first-floor walls had to be able to withstand the weight of the entire building. This caused the lower-floor walls and the foundation of the building to become gigantic. It was not uncommon then for walls in the basement of even modestly tall buildings to be twenty or more feet thick.

More important, perhaps, than wall thickness was the fact that elevators did not exist. People simply would not or could not walk up more than a few flights of stairs. These two factors then combined effectively to limit the height to which office buildings were built.

A state of equilibrium existed among building owners. Every owner of an office building realized about the same return or profit on his investment and therefore all office buildings were about the same height. The profit equilibrium was stable. An owner who might try to build a taller building quickly found that it produced no additional profits because no one would rent the space on the upper floors. Hence he could not realize a profit on the additional costs he had to pay for building sturdier foundations and walls. If he chose to build an office building with fewer floors than was customary at the time, his profits also fell because he did not have as much space to rent as his competitors. All the economic forces then were at work pushing him to build his buildings with about the same number of floors as his competitors. The system consisting of building owners, sellers of land to building owners, and renters of office space was in balance, or stable equilibrium.

At the end of the nineteenth century two things happened to change the nature of this equilibrium condition from stable to unstable. First, the cheap fabrication of steel beams made possible the construction of tall, light, steel-frame buildings. Second, the elevator was invented by Elisha Graves Otis and electrical motors became available for its operation. These inventions were not ignored by people who built and rented office buildings. New driving forces quickly altered the stable equilibrium condition.

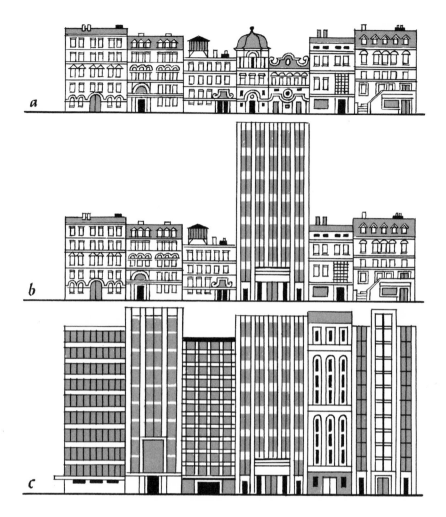

FIGURE 8
(a) Stable situation with all office buildings around five stories tall. (b) Unstable situation created when first high-rise building is erected. (c) A new stable situation develops relatively quickly when all low-rise buildings are torn down and are replaced by high-rise buildings.

The owner of the first skyscraper was able to realize a larger profit from a small plot of land. This quickly caused the price of land on which skyscrapers could be built to rise steeply. The owners of vacant land no longer sold at the price that had prevailed when buildings of only four or five stories were constructed on it. They held their land until a skyscraper builder came along who was willing to pay more for the same plot because his building would realize a larger return on his investment. As land prices increased, only a skyscraper was profitable and each new skyscraper created pressures for succeeding office buildings to be skyscrapers. New York City experienced a severe instability with land for office building quickly rising in price as building height zoomed from four or five floors to a typical height in excess of sixty floors in the space of thirty years or so.

This example illustrates both the concept of feedback and more importantly, perhaps, unstable equilibrium in an economic system. The negative feedback in such situations quickly changes an unstable economic situation into a stable one. The input or desired goal is a situation in which stable land prices prevail. That price which will bring just enough land to market to meet the demands of skyscraper builders can be called an equilibrium price or in general terms, an equilibrium point. (A goal of stable prices may be easier to understand when one considers how most people dislike economic surprises. It has been observed many times that people who buy milk regularly would rather pay thirty cents per quart day after day than ten cents per quart some days and fifty cents per quart on others.) The output is the actual price of land. In this example it is the competitive market which will bring these two prices together; that is, the marketplace set up by competition between land buyers and land sellers acts as the comparer and reducer in this closed-loop system. As in all the earlier examples the comparer and reducer act to reduce the difference between the input, or desired goal of stable prices, together with the output of actual prices.

Feedback is in the form of information about how much land is available for skyscrapers, and how much land is needed for skyscrapers. Before the invention of steel beams and elevators, land for buildings sold at a price that varied little from year to year. It was the disturbing force of these two inventions that drove the price of land to a new equilibrium price at which point the supply of land would just match the demand for it. Free markets will, almost without exception, set the price of a commodity so that its supply equals its demand. Another economic example will be considered in more detail in a later chapter.

Stability in Humans

One of the reasons that physicians and scientists have begun to think of the human nervous system as a feedback system is that it helps them to diagnose illnesses in parts of the system that might not be working as they should. Recently they have even been able to write down descriptions of some of the body's systems in mathematical terms just as an engineer might write down a mathematical description of a machine. The idea of examining parts of the human body in terms of feedback systems is not done just for fun or as a thinking game; it is done so that man might better understand how his body carries out the many different tasks it does each day. And then he can try and find ways to fix the feedback systems when they perform poorly or not at all.

Consider again the most familiar example of a feedback system— that comprised of the human eye, brain, and hand. In most people this feedback system works superbly well. Building a machine to pick up a pencil from a desk top would be a formidable task and yet you perform it routinely without even thinking consciously about doing it. It even operates well when you are subjected to disturbances such as being on an airplane or in an auto. In this system the output, or result, is the position of your hand while the input, or goal, is the position of the pencil. The person trying to pick up the pencil observes the difference between the position of his hand and the position of the pencil. By using his eyes as a comparer, he generates signals in the brain that will cause him to move his hand in such a way as to reduce the difference between it and the pencil. In addition to the main feedback through the eye, there is an additional feedback path directly from the muscular system to the brain: as the muscles are actuated, the individual senses the forces and how far his hand has moved. There is a third feedback loop based on the sense of touch that comes into play as soon as the hand touches the table top or pencil.

However, it is possible for feedback mechanisms in the human nervous system to fail. People who have such a failure are said to be suffering from an illness called *ataxia*. If they attempt to pick up a pencil from a table top their hand starts to oscillate or "hunt" as it approaches the pencil. In many cases the shaking of the hand is so violent that they are unable to pick up the pencil. Ataxia, then, is a disease that causes loss of control due to oscillation or instability within a portion of the central nervous system.

Sometimes a man-made feedback system that has been incorrectly designed will hunt to such a degree that it ultimately will destroy itself. The oscillations occur because the system overcorrects, just as in the case of a person suffering from ataxia. On occasion, guidance systems on space probes have gone into uncontrolled oscillations. First the control system calls for a one-degree change in attitude to the left to achieve a desired course, and in making the correction it overshoots the correct course by two degrees. Then the control system calls for an overcorrective course change of three degrees to the right because of the severe course error at the present heading. This new correction puts it further off course and in need of more correction than before. If the system continues to order such corrections it will very soon cause the probe to shake itself to pieces because of the ever increasing course changes. Whether or not a control system for any feedback device will produce unstable oscillations, or uncontrolled "hunting," as it operates is one of the first things a designer must check.

Though systems designers can usually take those steps which are required to produce stably operating devices, the actions necessary to cure unstable oscillations in people with ataxia have not been so easily attained. Can such diseases be cured? The answer to that question is not yet known, but it is safe to say that the more that is learned about feedback in general, the more likely it is that a cure for ataxia and similar diseases will be found.

4 The Origins of Feedback

FEEDBACK is a universal process that can be seen in a great variety of phenomena, from the population cycles of predatory animals to the fluctuations of the stock market; every animal employs self-regulation to ensure its existence and stability. Considering how universal this phenomenon is, it seems curious that a scientific study of the idea of feedback control came so late in the development of science and technology.

The word "feedback" itself is a recent invention. It was coined by pioneers in radio around the beginning of this century. However, wide-scale exploration and exploitation of the feedback principle did not begin until the 1940's. It received its main impetus from the work of the late Norbert Wiener and his colleagues.

For the most part in the last couple of centuries, science has led technology. That is, a fundamental law of nature is discovered, then someone comes along and applies that law in one or more useful inventions. However, in this case feedback control in machines is an example of technology giving birth to a science. Application of the feedback principle had its beginnings in simple machines and instruments, some of them going back two thousand years or more. More recent examples of feedback control are the flyball governor and the thermostat.

Though many ancient inventions that used feedback control were well thought out and were fairly complicated, feedback control as an abstract idea did not receive much attention until the 1930's when both biologists and economists began to note striking parallels between their own subjects and the feedback control devices of engineers. They recognized that the control processes in both living organisms and in economic systems showed the same cyclic structure of cause and effect. It soon became evident that the idea of feeding back a portion of the output of a system could be a versatile and powerful tool for investigating many forms of dynamic behavior. Today this idea is not only incorporated into thousands of different types of devices but also recognized as an important unifying idea in science.

The purpose of all earlier feedback devices was exactly the same as the purpose of all modern feedback devices—to carry out a command automatically in order to maintain the output at a desired level, as specified by the input in spite of any interference by unpredictable disturbances. The command signal may either be constant or changing with time.

The origin and main lines of development of the feedback concept are illustrated by three devices: the ancient water clock, the thermostat, and the mechanism used to control windmills.

The Ancient Water Clock

So far as is known the earliest feedback control device was a water clock invented in the third century B.C. by a Greek inventor named Ktesibios. It is believed that he was associated with a well-known museum that was then the main cultural center of the Mediterranean world and attracted some of Greece's best scholars. Ktesibios' own descriptions of his inventions are now lost, but fortunately an account of them is preserved in a book called *De Architectura* written by the Roman architect and engineer Vitruvius.

Though Ktesibios did not invent the water clock (it was probably invented by the Chinese about one thousand years earlier) he did add to the Chinese's original design a feature that had not been seen before—self-correction. In the water clock invented by Ktesibios the passage of time was measured by means of a slow trickle of water, flowing at a constant rate into a tank where a small indicator floating on the water told the time as the water level in the tank rose (Figure 9). The problem of maintaining the trickle at a constant rate was solved by inventing a device that in many respects is similar to the modern automobile carburetor. Placed between the source of the water supply and the receiving tank was a float valve.

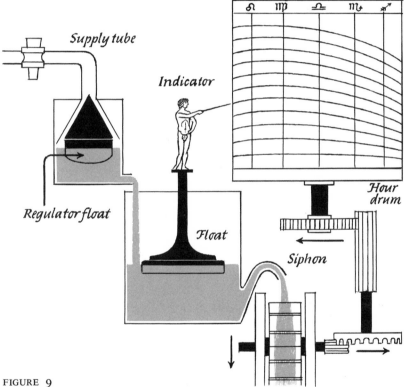

FIGURE 9
One of the first self-regulating water clocks believed to be invented by Ktesibios. If the volume flow rate of water entering the clock is reduced, the regulator float sits lower in its chamber, thus allowing more water to enter from the supply tube. When the flow rate increases, just the opposite occurs. These actions keep a constant rate of water entering the big chamber.

If the flow of water should increase for some reason, this would cause the float valve to rise in the small-float tank and reduce the flow of supply water; if the flow of water from the source should fall, then the float would ride lower in its tank and thus allow an increase in the flow of supply water. The adjustment in the flow required no human intervention and was completely self-regulating. The output was the speed at which the float rose in the tank; the input or desired goal was a uniform rate of rising of the float since the clock was to indicate a constant rate of passage of time regardless of water flow conditions. The comparer and reducer were combined in the regulator float as it sat in its small chamber.

A careful study of the history of science from the time of Ktesibios until the Middle Ages reveals numerous other applications of automatic regulation as carried out by float-type devices. Three centuries after Ktesibios' invention, Hero of Alexandria, a great writer, mathematician, and inventor, described numerous devices demonstrating the principles of feedback control. Investigation of Islamic technology in the period preceding the Middle Ages has revealed ingenious use of the float valve to achieve automatic control in differently designed water clocks and other devices.

The Thermostat

Thermostats are used to maintain objects, rooms, or even whole houses at a fixed temperature. Modern man has found thousands of uses for thermostats and has now designed many clever devices that will maintain a constant temperature. However, the thermostat is a mechanism that man has come upon much more recently than the water clock.

Cornelius Drebbel, a Dutch engineer who had migrated to England in the seventeenth century, is thought to have first invented the thermostat. Drebbel was not a great believer in writing things down and no doubt his work would be much better known if he had committed his numerous inventions to writing. Fortunately, Francis Bacon, another well-known scientist of Drebbel's day, has left a description of Drebbel's thermostat. He claims that Drebbel devised his temperature regulator only as an aid to his main work, which was alchemy. (Alchemy was a half-magic, half-scientific activity that had as its goals, the changing of metals such as lead into gold, the discovery of a medicine that would cure all illnesses, and the preparation of a potion that would provide eternal life.) Drebbel apparently believed that he could transmute base metals to gold if he could keep the temperature of the process constant for a long time.

FIGURE 10

A drawing made by Drebbel's grandson of the thermostat invented by Drebbel in the seventeenth century. A fire is located in the region marked *a-a* which gives off smoke which passes by a water-jacketed incubator box shown in dotted lines and then goes on to escape out of the opening marked *d*. A small glass vessel filled with alcohol (*c*) is inserted into the incubator box and is sealed with mercury contained in the U-shaped tube at the right. When the temperature increases, the volume of the evaporating alcohol increases, forcing the mercury to rise in the right leg of the tube. This raises a float (*b*) and through a linkage (*f*) closes a damper (*e*), thereby reducing the fire and ultimately lowering the temperature in the incubator.

Drebbel built a box with a fire in the bottom and above this placed a small container with a U-shaped neck topped by mercury; in the container he placed air or alcohol. As the temperature in the box rose, the increased pressure exerted by the air or alcohol vapor pushed up the mercury, which in turn pushed up a rod; the mechanical force of the rod was applied to close a damper which would throttle down the fire. On the other hand if the temperature in the box fell below the desired level, the gas pressure was reduced, the mercury dropped, and the mechanical linkage opened the damper which would cause the fire to become stronger.

This device was used by Drebbel not only for smelting experiments but also to maintain an even temperature in incubators for hatching chicks. His regulator apparently impressed some of the more notable members of the Royal Society of London favorably, including Robert Boyle and Christopher Wren. Drebbel's grandson prepared a detailed description of his grandfather's invention, including a sketch that showed how it worked (Figure 10).

More than two centuries passed before any real interest in temperature regulation by feedback again attracted attention in the engineering community. A Parisian inventor named Bonnemain, who had read some of the descriptions of Drebbel's and others' work on temperature control, built a temperature regulator himself for which he was able to get a French patent. He succeeded in using his temperature controller in an incubator at a large farm that supplied chickens to the royal court and the Paris markets. Bonnemain's apparatus was superior to earlier temperature regulators as it had a far more sensitive temperature feeler and several other improvements in design. Bonnemain refrained from showing the details of his apparatus to the world at large until he was over eighty. Apparently he wanted to keep the details of his thermostat secret because of the edge this gave him in manufacturing and selling it. In 1824 he finally published a description of his system of temperature control.

After the leading technical magazines in Britain and Germany published translations of the details of Bonnemain's temperature regulator, discussion of temperature regulation soon found its way into encyclopedias. The author of one of these, the Scottish chemist Andrew Ure, coined the term "thermostat" in his *Dictionary of Arts, Manufactures, and Mines,* which in 1839 described Bonnemain's regulator and one that Ure himself had designed.

Many modern thermostats are remarkably similar to the device invented by Bonnemain. They use a small bellows filled with a gas which expands or contracts as its temperature rises or falls. As the gas in the bellows cools, it shrinks and closes an electrical switch which turns on a furnace or other source of heat (Figure 11). Because they are so simple, thermostats generally last for years without need of maintenance or adjustment.

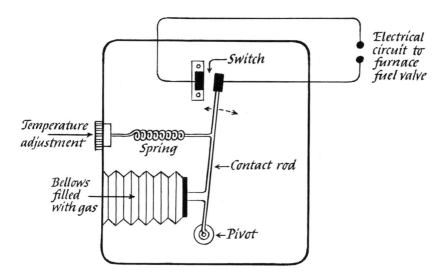

FIGURE 11
Diagram of a simple thermostat. The bellows filled with gas serves as a thermometer in this device; as the temperature of the room where the thermostat is located drops, the gas in the bellows shrinks, causing the bellows to move the contact rod to the left. This closes a switch that electrically opens the fuel valve to the furnace. Thus the temperature in the room will increase until the desired temperature is reached, at which point the thermostat acts to cut off the fuel flow to the furnace.

Windmill Control

The eighteenth century in England and Scotland was a time of great advance in the mechanical arts. A skillful group of millwrights were able to combine their craft skills with the beginning of a scientific attitude. In the eighteenth and nineteenth centuries many men who began their careers as millwrights became some of Britain's most famous mechanical engineers. Several of them made important contributions to the application of the feedback principle.

In order for a windmill to make the most effective use of the wind it must be kept facing into the wind. Furthermore, since in most places the wind frequently shifts direction, the problem becomes one of adjusting the direction of the windmill into the wind even though the wind might be changing its direction every few minutes. Prior to Edmund Lee's invention in 1745, changing the direction that the windmill faced was left up to the miller or his helpers. Lee's invention was the first truly automatic servo—a closed-loop device that can follow a *changing* command signal. It consisted of a fantail, which is a small windmill mounted at right angles to the main wheel. By means of a set of gears and a linkage, the fantail is able to rotate the movable cap to which the main wheel is attached. When the main wheel squarely faces the wind, the fantail is at right angles to the wind. In this position the fantail is parallel to the direction of the wind and does not rotate. Whenever the wind shifts its direction so that the main wheel no longer faces it squarely, the wind will strike the tail wheel, causing it to rotate slowly and turn the mill cap until the fantail is again parallel to the wind and the main wheel faces into it (Figure 12). Thus the fantail becomes the comparer and reducer which senses the desired direction and compares it with the actual direction. It then tries to reduce the difference between the two directions.

Lee did not stop with just controlling the direction of the main wheel. He also addressed himself to the problem of the speed at which the mill should turn. The problem arises because not only does the wind's direction change frequently but so does its speed.

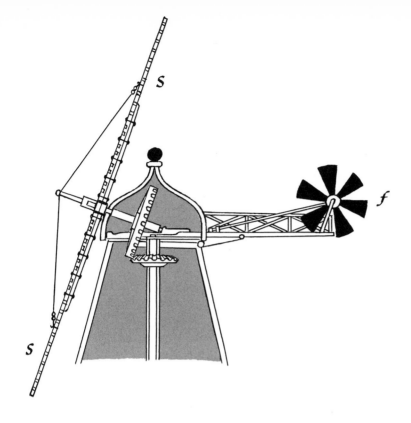

FIGURE 12
Adapted from an original drawing for a fantail for a windmill. The fantail
(*f*) will locate itself at right angles to the wind by causing the top of the
windmill to rotate. In so doing the main blades of the windmill (*s-s*) will face
squarely into the wind, where they will deliver the most energy to the mill.
If the wind shifts direction, the fantail will turn the mill again until maximum
power is achieved.

Regulation of how fast the millstone would turn was needed to protect the millstones from excessive wear and to produce flour of uni-
formly fine quality. Lee solved this problem by allowing the sails to
pivot around the arms that held them. The sails were connected
to counterweights that caused their leading edge to pitch forward in
moderate winds. If the wind's velocity increased excessively, the
force on the sails became greater than the counterweight and the tilt
of the sails was adjusted to reduce the power delivered to the millstone and thus its velocity. In this way the mill acted to shut
itself down in very high winds. This arrangement, of course, would

FIGURE 13

Adapted from an original drawing of a feedback device for setting the sails properly on a windmill. When the forces on the windmill blades (g) became greater than the force exerted by the counterweight (h), then the tilt of the blades was adjusted through the linkage (e-d-c-f) to reduce the power delivered to the millstone. When the wind velocity fell, the mechanism sensed the change and again allowed the blades to take more of a bite out of the wind.

sense when the wind's speed had subsided and adjust the pitch on the sails accordingly (Figure 13).

The millwright occupied a unique position—as a craftsman he combined the skills of a carpenter, mechanic, and blacksmith—but at the same time he was faced with solving theoretical problems associated with wind and water wheels. His profession caused him to travel from mill to mill and thus he was exposed to the latest developments in his art as well as to new theoretical ideas. He thus became the bridge between the traditional craftsman and the engineer trained in the physical sciences.

Both the strengths and weaknesses of the millwrights were reflected in their regulating devices. They grasped new ideas with enthusiasm and imagination but they were not always able to bring them to a stage of completion. It was only when the idea of feedback was applied to the steam engine, which was the main driving force for the Industrial Revolution, that this important idea became truly effective.

The Flyball Governor

One of the devices used to control the speed of millstones and to set the sails on windmills was a so-called centrifugal pendulum. It consisted of two metal balls which were free to swing out if the object they were attached to was rotated. They move out because of the centrifugal force that is created in any object that is put into rotation. You can observe such a force yourself simply by tying a piece of wood on the end of a piece of string and then swinging it in a circle about your head. The faster you swing it, the further out the piece of wood will move.

The idea of the centrifugal pendulum was greeted with enthusiasm by the pioneers who were working with the developing technology of the steam engine which was beginning to emerge late in the eighteenth century. James Watt and his partner Matthew Boulton were building a large mill (later to be named Albion Mill) where Watt was planning to demonstrate the virtues of his new rotary engine. This engine called for an improved means of regulation; there did not appear to be any easy way to adapt existing devices to control the continuously operating rotary engine.

To supervise the construction of Albion Mill, Watt and Boulton hired John Rennie, then a young man of twenty-three. He had just completed his apprenticeship under a noted Scottish millwright.

When Boulton was visiting Albion Mill in 1788, he found that a device had been installed, presumably by Rennie, to separate the millstones if they started to rotate too quickly. The device used a centrifugal pendulum to sense the mill's speed. Boulton quickly sent off an enthusiastic and detailed description of it to Watt. The timing was perfect. Watt and his co-workers designed a "centrifugal speed regulator" and around the end of the year the first governor was installed. The picture of Watt's flyball governor was to become perhaps the most familiar symbol of Europe's transition from an agricultural society to an industrial one (Figure 14).

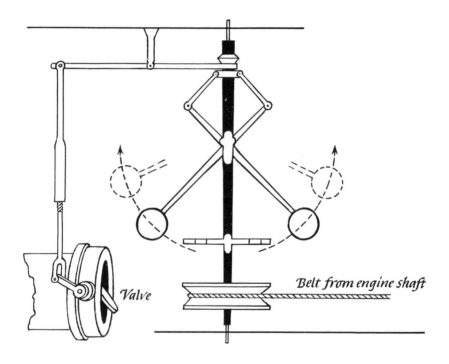

FIGURE 14

The flyball governor that was used on early steam engines. As the speed of the engine increased, centrifugal forces pushed out the whirling balls, which through a linkage acted to reduce the supply of fuel to the engine and thus slow it down. Just the opposite happened when the engine slowed to a speed less than which was desired.

Because he believed that his governor was merely an adaptation of the centrifugal pendulum, Watt did not take out a patent on it. Instead, he and Boulton asked their customers to hide the governor from view in order that competitors could not learn how control of the steam engines was being achieved. Of course, the device soon became known anyway and every steam engine soon had one. The flyball governor demonstrated the action of feedback control more widely and forcefully than words could have done. It was not long before the governor entered the textbooks and handbooks of engineering, and inventors began to develop feedback devices in other areas of technology.

But it was almost a hundred years before scientists began to notice that animal populations also could be studied on the basis of feedback systems. Another fifty years after that had to pass before physiologists recognized the feedback loops that make up the autonomic nervous system that controls body temperature, pulse rate, and the like in all living organisms. The next few chapters will consider feedback as it occurs in living systems—first in a single complex organism—the human body—and then in groups of living organisms, both animal and human.

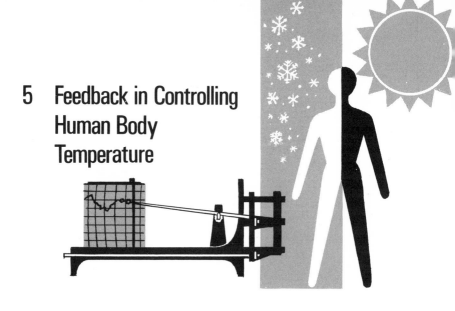

5 Feedback in Controlling Human Body Temperature

THE HUMAN BODY contains within it a number of feedback control loops which can be very instructive in understanding the principles of self-regulating systems. Consider the problem of maintaining the human body at an internal temperature of 98.6°F. The problem is particularly interesting because it illustrates a case where feedback is used to ensure a highly accurate output even when the system is subject to large external disturbances.

Human body temperature must be controlled very accurately as cells of the adult nervous system are damaged if the temperature rises and stays above 108°F for a prolonged period of time. Even a three-degree temperature drop results in a greatly reduced enzyme activity in the body. For this reason a normally healthy individual will have a body temperature within one degree above or below 98.6°F.

Human beings may be subjected to air temperatures as high as 120°F or as low as −60°F, although such extreme temperatures are the exception rather than the rule. Results of research done to protect people who work in the Arctic or the Antarctic indicate that the wind can be just as important a factor as the temperature as far as the body is concerned in maintaining its own temperature. For example, a temperature reading of

−20°F accompanied by a 25 mile-per-hour wind is equivalent to a temperature of −74°F; at such a temperature exposed flesh will freeze in less than one minute. It should be easy to recognize the similarity between the body's system and the example considered previously of controlling the temperature inside a house. In the human body, however, nature provides several different means by which temperature is measured as well as several different sources of heat rather than the single thermostat and furnace commonly found in a house.

The three sources of heat used to influence body temperature are the core, the skeletal muscles, and the skin. As the body attempts to maintain a constant core temperature, it may use only one of these three mechanisms; but under extreme conditions it may attempt to use all three simultaneously.

Energy is released in the *core* by means of a chemical reaction which oxidizes, or slowly burns, fat and other primary foods. Your body takes the food you eat and breaks it down into simpler chemical compounds which are then stored in the body's cells. When you breathe you take in oxygen which is then used to burn (at body temperature, of course) the simpler chemicals which were derived from your food. The oxidation or burning of the chemicals derived from food that is eaten results in the release of carbon dioxide to the blood along with heat energy. This process of turning food into energy is called metabolism and is fundamental to all living things. How fast food is turned into energy is called the metabolic rate. During exercise or times of great stress or anxiety the net consumption of oxygen will increase as breathing becomes deeper and the heart rate increases to speed up the metabolic process. Temperature control, however, is achieved by control of the metabolic rate through a reducer known as the endocrine gland, which receives electrical signals from the brain (the comparer) and orders the metabolic rate to speed up or slow down.

Muscles can also be a source of heat. If the skin detects a sharp drop in outside temperature, electrical signals are sometimes transmitted from the brain to the muscles ordering the body to shiver.

Shivering causes the muscles which are next to each other to operate in an uncontrolled fashion, with the result that there is very little useful work done as most of the energy of the muscles is turned into heat. It is also possible for there to be some metabolization taking place in the muscles, thereby generating heat.

The *skin* is used to cause changes in the internal temperature in two ways. One way is to control the amount of blood allowed to come close to the surface of the skin—this is called the vasomotor effect. For example, if you take a cold shower the blood vessels near the surface of your skin will contract and restrict the amount of warm blood they carry. A second method of temperature control that the skin uses is sweating. Sweating encourages the loss of heat from the skin; it does this by using some of the body's internal energy that is being delivered to the skin's surface where it evaporates water. The evaporation of the water produces a cooling sensation on the skin. Sweating is an especially important method of cooling for humans when they are engaged in heavy labor or exercise, or when the surrounding temperature is higher than the body temperature. (Because dogs have very few sweat glands, they pant to increase evaporation from their tongue and mouth.)

To summarize, human beings use their core, skeletal muscles, and skin in four primary ways to exert control over their internal body temperature: variation of the metabolic rate, shivering, varying blood flow to the skin, and sweating. Each of these four types of control is actuated by signals from that region of the brain which determines temperature control. This part of the brain is given information (by way of tiny electrical impulses) about the internal body and skin temperature. This information is produced by nerve sensors which respond to temperature and to temperature changes.

With this background it is now possible to construct a loop diagram illustrating the components of the feedback systems and showing the various feedback paths. The main output of the system is the actual internal body temperature or core temperature. A secondary output is the skin temperature. The main input to the system is the desired core temperature (normally 98.6°F). In

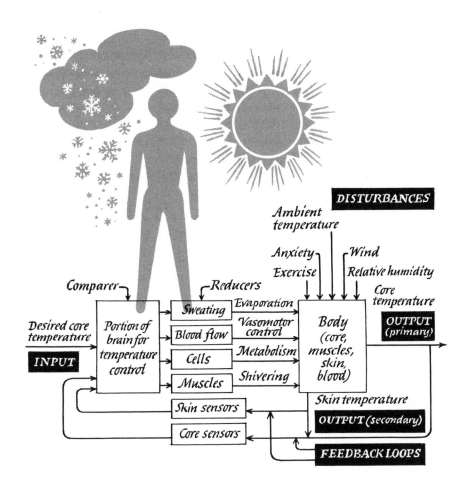

FIGURE 15
The major feedback loops that control the body's core temperature and skin temperature. A portion of the brain charged with maintaining a constant core temperature is the comparer in this example. Note the multiple feedback loops and two outputs.

addition to the primary input signal to the body there are also possible disturbing forces which can influence core and skin temperature. These disturbances are exercise, anxiety, outside air temperature, wind, and relative humidity.

In the case of many illnesses, the input signal is changed to produce a fever. It is not completely clear how this change is produced. In certain illnesses the temperature response of the patient exhibits instabilities. For example, people with a common flu infection frequently suffer periods of fever followed by periods of chills. During the fever portion of such illnesses the body may be working extremely hard to reduce body temperature and in fact may be so successful that the feedback system lowers temperature to a value below normal, producing chills and shivering. The control system then corrects itself and again overshoots the mark, producing fever. Such big swings in body temperature have a tendency to make the patient feel very weak.

On the diagram, you will see that one of the outputs shown is the skin temperature. This is a particularly important output because it can alert the control system to a sudden shift of temperature before the core temperature has a chance to feel it. For example, if you step outside on a cold day without a coat on, your skin will immediately note the change and start constricting your blood vessels in order to keep inside as much of the body's energy as possible. In this way the body has a head start on maintaining its normal core temperature without waiting for the chilling effect of the outside air temperature to reach the core.

The block diagram presented in Figure 15 is believed to include the primary elements and signals which make up the temperature control system of the human being. What is the value of diagrams of this sort? How do such descriptions aid researchers in learning about the operation of a system? How might the model help in developing improved medical procedures?

To answer these questions is difficult, but in general the usefulness of feedback models comes about when more detailed information is available on each of the blocks in the feedback diagram. Certain comments about the uses of the study of feedback systems can be made just from the block diagrams.

It should be clear from the discussion of those factors which can influence body temperature how important it is for the body to have a feedback mechanism. The internal or core temperature of the body changes very little (typically less than one degree) even when subjected to radically different levels of exercise, stress, and the environmental effects of temperature, wind, and relative humidity. These disturbing forces are not part of the feedback loop but usually act on the system in such a way as to try to push the output away from the desired input. That is, stepping outside when it is −20°F will have a tendency to lower both the skin temperature and the internal body temperature. But the body's feedback mechanisms will perceive this undesired push away from the goal of a constant body temperature of 98.6°F and will cause the four processes described earlier (increased metabolization, shivering, and so on) to counteract the effect of the −20°F environment.

It is also interesting to note that the core temperature is almost completely independent of changes in the body's properties. For example, changing one's clothes or an injury do not alter the core temperature. Neither do major disturbances such as amputation by surgery alter the performance of the feedback system.

The model illustrated also points out those areas which need more scientific research to gain a better understanding of how the system works; it can be used to raise basic but important questions. For example, how is the input signal (normal body temperature) changed (if indeed it is) to cause a fever when a patient is ill? Further study of this question may show that the input is not changed at all, but rather that during illness the system loses effective control of the temperature so that it rises. If this is true, perhaps a major effort should be made to hold down the temperature of an ill patient. But if research revealed that fever is an important contributor to the body's power to resist disease, perhaps physicians should not indiscriminately use drugs which reduce the temperature—at least not until the patient's temperature approaches the danger range.

Like almost all models of real physical systems the model presented here is approximate. As more and more is learned about the mechanism of temperature control in the human body, interactions not represented in this model will be discovered. Then, as the model is revised, new experiments will be suggested that will further improve understanding of the system. Thus, the feedback model can serve as a key tool in the development of scientific understanding.

Finally, man can learn to use nature's millions of years of experience to improve upon the devices he builds. The close similarity between the human temperature control system and a house heating system, has already been noted. Some large heating and cooling systems now use outside thermometers as well as inside ones to provide information to the temperature control system of the building. In this way the outside air temperature, like your skin temperature, can be used to anticipate demands that will be made on a building's heating and cooling system.

As interesting as the control loops that make up the autonomic nervous system are, some scientists believe that feedback plays its most important role in what goes on in man's mind. That is, does not the process of learning itself simply demonstrate feedback operating in a very basic way? It is this important question that will be considered next.

6 Feedback as an Aid to Learning

LEARNING has been defined by psychologists as a process which brings about a change in a person's way of understanding or responding as a result of experience. It is the phrase "as a result of experience" which introduces the idea of feedback into the learning process. It is not very difficult to draw the feedback loop for any learning process, even though how people or animals actually learn is far from being well understood.

That feedback is necessary to the learning process has been recognized for hundreds, if not thousands, of years. Figure 16 illustrates an example from the fifteenth century. Knights in the time of Charlemagne learned jousting skills with a teaching machine called a quintain. The student-knight charges at a wooden figure on a pivot. If he strikes it squarely in the middle of the shield it will fall over. If he strikes it off-center the teaching machine will supply feedback information to the student-knight, informing him that he has done poorly and that he should strive to be more accurate in the placing of his lance. The feedback information will come in the form of a blow from a club held by the quintain. Learning apparently took place very quickly when this effective means of feedback was supplied to the student.

A Little About Learning

Some psychologists believe that any animal learns in two different ways; one way is called classical learning (or conditioning) and the other is called instrumental learning. Understanding the difference between these two different forms of learning is necessary in order to see how feedback enters into the learning process.

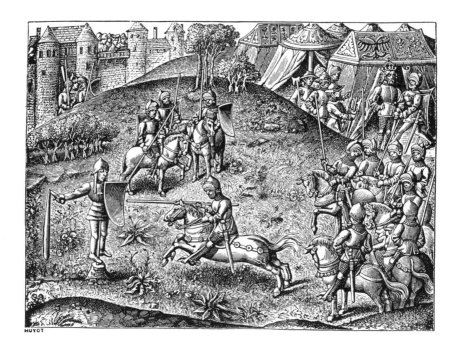

FIGURE 16

It has long been recognized that immediate feedback is an important aid to learning. Jousting was taught in Charlemagne's time with this teaching machine called a quintain. A knight charges at the wooden figure, and if he strikes it squarely in the middle of the shield, it will fall over. A hit made off-center, however, will cause the club held by the figure to swing around and firmly apply feedback to the student knight. (Fifteenth century woodcut from *Chronique de Charlemagne,* International Science and Technology Print Room, New York City Public Library.)

Classical learning is something that all of us have experienced whether we realized it at the time or not. Suppose that you had never gone swimming in a really cold mountain lake; a lake that not too many weeks before had been filled with the runoff from recently melted snow. The first time you saw the lake you were impressed with its beauty. But then you dove in and learned how cold it really was and you started to shiver as your body responded with all of its defense mechanisms for protecting itself against suddenly falling external temperatures as was described in the previous chapter. After you had gone swimming a couple of times in this lake, the shivering and reduced blood flow to the skin would begin even before you got into the water. It would begin when you got down to the lake and just took a look at that icy blue water. You have been classically conditioned to shiver at the sight of a lake full of cold water.

Instrumental learning is much less limited in its usefulness since it does not require a specific stimulus (the ice-cold lake water in the previous example) to get a response. Suppose you want to learn to throw darts from, say, a distance of ten feet. There is no stimulus anyone can offer you—neither money, honors, nor praise—that will cause you to hit the bull's-eye the first few times that you try. But by the process of trial and error you can learn to coordinate your muscles and your eyes. At first your throws will miss by a wide margin but if you keep on trying, the dart will begin to come closer and closer to the center of the target. You are not quite sure of precisely what it is you are doing to improve; you cannot say for certain which muscles are contracting and which are relaxing. But as the darts hit closer to the target center they act as a reward which is likely to keep you trying. At last you hit the bull's-eye. This trial-and-error process is called instrumental learning. The term instrumental comes from the fact that the response of the learner is "instrumental" in obtaining a change in its relation to the environment. You learn to throw darts (your response) and you are rewarded by becoming a good dart thrower. Your trial-and-error efforts were instrumental in gaining you this reward.

Closing the Loop Again

Now suppose that you put on a blindfold and try to throw darts. The task becomes very difficult if not impossible. It becomes difficult because an essential element is missing from your system. That element is, of course, feedback. In this case the feedback is the sight of the dart hitting the target. It is this mental picture that your brain uses to send signals to your arm muscles which causes you to make the corrections needed in order to make bull's-eyes. Now if a friend were watching you throw while you were blindfolded, he could tell you if you were facing in the right direction, where your throws were landing, and whether you should be throwing to the right or to the left, up or down. Using this second-hand information, you might learn to throw darts blindfolded. But second-hand feedback is not nearly as good for learning as first-hand information gathered by your own eyes.

Psychologists have now come to realize just how important feedback is to the learning process. It is often assumed—in the classroom, office, and shop—that one person can effectively transmit information to another without being aware of the latter's reaction. In fact, communication is only effective if it is a two-way street. If A is to be efficient in transmitting information to B, he must know whether B is receiving and understanding the information. That is, there must be feedback from B to A.

A research study has been carried out on the necessity of feedback to the learning process. Different levels of feedback were tested on four sets of students:

(1) *Zero feedback.* Teachers were completely separated from students. The teacher could not see or hear the students.

(2) *Visible audience.* The students could not speak to the instructor but the instructor could see the students.

(3) *Yes or no feedback.* Students were permitted to respond with a "yes" or "no" to questions from the instructors.

(4) *Free feedback.* Students were permitted to interrupt
at any time to ask questions of the instructor.

The instructor was trying to teach the students to draw abstract
geometric patterns; the students had to draw the patterns based on
what the instructor told them. Accuracy of the drawings increased
steadily from zero feedback to the free feedback condition. But the
time required for the student to learn to draw the patterns also
increased as the level of feedback increased. Another interesting
result was that free feedback generated confidence in the students,
whereas the condition of zero feedback tended to make the students
angry.

In 1926, Sydney L. Pressey, a professor of psychology, suggested a way to use instrumental conditioning in a systematic way
in the teaching process. He proposed that machines be used to teach
as well as to test and score results. Pressey seems to have been the
first expert to emphasize the importance of immediate feedback in
learning and to propose a system in which each student could move
at his own pace.

Pressey's ideas were largely ignored for more than thirty years,
except for a few experimental programs carried out in the military
services. During the past ten years, however, there has been a tre-
mendous increase in interest in the use of self-instructional ma-
terials.

The essence of any programmed instruction is that new material
is presented in steps, sometimes called "frames"; each section of
material that is to be learned ends with an item which the student
must complete or answer correctly before proceeding to the next bit
of information. Thus a student knows exactly how he is doing at all
times, and if he has missed a point or misinterpreted the material,
he finds out about it at once. Of course, it is the informing of the
student about how well or poorly he is doing that constitutes the
feedback in the system. The student is never permitted to go on to
the next frame without first having mastered the information already presented.

There is considerable debate about whether or not programmed material is a good way to teach. There is no argument about the feature that allows a student to proceed at his own pace rather than the pace set by the teacher. Teachers have always known that different students require different amounts of time to cover the same material, but they have never been able to arrange the classroom situation that would permit thirty or forty students to proceed each at his or her own pace.

How much a student will learn from programmed instruction depends upon how well the program is written. It turns out that writing a good program is an extremely difficult task. Ideally, a teaching-machine program should be so constructed that the number of errors that the student makes while proceeding through the program is kept very small. But how is the person who is constructing a teaching-machine program to know whether the program can be navigated with little error? The only way to answer this question is to let students try it out and see how many errors they make. What the program-writer gets from this experience is, of course, feedback. He finds out what part of the program seems to be doing the poorest job, revises it, and then lets a new group of students go through the program. In this case feedback is teaching the programmer to write good programs.

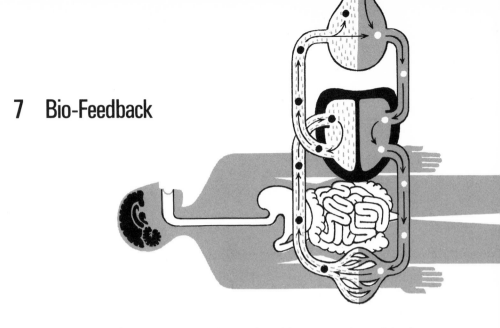

7 Bio-Feedback

CHAPTER 5 discussed how the body utilizes a number of feedback loops in a way that man apparently has no control over. That is, man's control systems seem to maintain a suitable internal body temperature and skin temperature without any conscious effort on his part. Similar feedback loops maintain an appropriate blood pressure, muscle tension, and the like, also without any conscious effort. These feedback loops make up what is known as the involuntary or *autonomic nervous system,* and for years researchers assumed that man had no way to intervene in its operation.

Efforts by a number of scientists have now completely changed that view of man's abilities. They believe that with the aid of electronics the mind of man can gain new powers over his body. Some believe their research—which is still in its infancy—may one day have a dramatic impact on medicine, psychology, industry, and education.

An individual's mind is in some ways like the blindfolded dart thrower where the involuntary nervous system is concerned. Researchers are now trying to find new ways to remove the blindfold, or to substitute second-hand feedback information to the mind. In this way it might learn to use information it now, for some reason, dismisses or never understands. The key to bringing the autonomic nervous system under control is a teaching method known as bio-feedback training. Such training allows an individual to observe the internal conditions and rhythms of his body. The individual is able to observe his condition by means of instruments. Once he can watch his heart rate, or listen to his brain waves, an ordinary man can learn to influence them at will without any noticeable effort. This control implies that many heart, circulatory, and other diseases may one day be treated without drugs or surgery. Furthermore, some psychologists believe feedback training may also open new doors in the mind, improving memory, learning ability, and creativity.

Here is an example of how bio-feedback can be used instrumentally to condition internal organs of the body by letting the mind know what those organs are up to. Dr. Bernard T. Engel, a psychologist, and Dr. Eugene Bleecker, a heart specialist, have had considerable success in teaching a group of their patients who suffer from a dangerous irregularity in their heartbeat to control the speed of their heart by utilizing mental efforts only. The changes in the rate of heartbeat that their patients could produce were not insignificant; they found that they could speed up or slow down their hearts by as much as twenty percent, which is sixteen beats per minute for a heart that beats normally at eighty beats per minute.

The key to teaching someone to control a part of his autonomic nervous system such as his heart rate lies in providing him with a stream of up-to-the minute accurate information about his heart rate—that is, he must receive feedback. While the patient lies quietly in a small room, the metal contacts of an electrocardiograph (EKG) are attached to his chest, wrists, and ankles—his pulse points. These contacts pick up the faint electrical signals given out by the heart and normally would be fed into a recorder device which uses a pen to trace on to a long roll of paper a picture of the frequency and shape of the patient's heartbeat. Such heartbeat recordings provide feedback of a sort to the patient and his doctor but it comes way too late for the patient to use it in any way that will help him to learn to control his own heartbeat.

Instead of feeding the faint signals from the EKG into a recorder, Doctors Engel and Bleecker direct them into a small specially programmed computer where they are almost instantaneously analyzed.

FIGURE 17
A research volunteer at New York University learns to relax by listening to the clicks she hears in the earphones every time her forehead muscle contracts. Whenever a muscle fiber contracts, it generates a very small electric voltage which is sensed by the metal contacts seen attached to the subject's forehead. These tiny voltages (around a millionth of a volt) are fed into a special amplifier which filters out those signals which are irrelevant and then multiplies the muscle voltages enough to create a clicking sound in the earphones. In that way the subject hears even the most insignificant changes in her muscle tension. Photograph courtesy of Lew Merrim from Monkmeyer Press Photo Service.

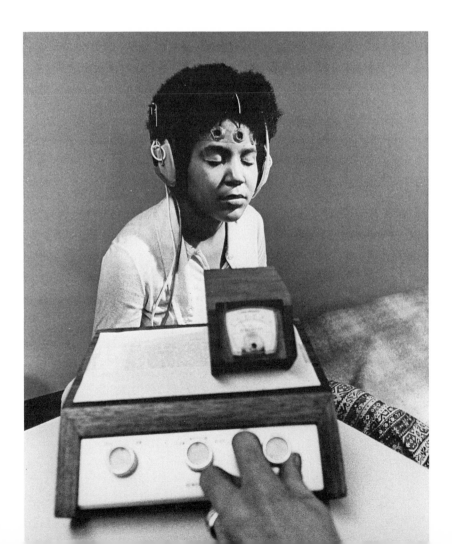

The output of the computer is fed into a small box sitting on a table that looks much like an ordinary traffic light with a red, yellow, and green light displayed on it. It is this simple three-colored light signal that provides the necessary feedback information to the patient.

By watching which color light is on, the patient is informed as to what his heart rate is doing. Thus a yellow light tells him that his heart is beating faster while a green light tells him that his heart rate is slowing. In order to make it easy for the patient to judge how well he is doing a small meter located next to the light tells him what percent of the time the yellow light is remaining lit if the patient is trying to achieve a faster heart rate. By gaining control of the lights, the patient ultimately gains control over his heart rate. The patient is instructed not to use muscular exertion at all in his efforts to control his heartbeat; only the mind is considered to be a legitimate tool of control in bio-feedback training.

The next phase in the training teaches the patient to learn to slow his heart down by using the red light as a guide. He later must learn to keep his heart beating within narrow limits, using the three lights to complete the feedback loop that runs between his brain and his heart. The red light tells him when his heart is beating too fast, the green light means too slow, and the yellow light means his heart rate is normal. Researchers have found that this self-willed change in heart rate tends to even out irregular beats.

The final phase of the program involves withdrawal of the bio-feedback light signals. After completing this stage of training, the patient finds he can purposely change the rate at which his heart beats in the desired direction without the use of the light signals. Many researchers are excited by the idea that Dr. Engel's patients could retain what they have learned without artificial feedback. They feel that maybe man can discover unknown feedback loops within himself that will allow him, after some basic training, to monitor and control throughout life various systems in his body now controlled by the involuntary nervous system.

Experiments have been carried out which try to teach people with high blood pressure to reduce it by thinking it down—using artificial feedback information at first. Also work is being done to see if subjects can learn to control the contractions of their intestinal tracts.

Some of the most exciting research, however, is on feedback training with brain waves—the electrical activity that all normally alert human brains produce from the moment of birth. There are four major types of waves produced by the brain—alpha, beta, delta, and theta. They are produced by billions of tiny electrical pulses that surge through the brain as it does the very complicated job of running a human body. A high rate of production of alpha waves is often associated with a state of peak mental and physical performance—relaxed and yet extremely sensitive and alert. By contrast, creative moods and problem solving tend to produce theta waves and waves which are combinations of alpha and theta waves. Intense concentration and a worried state seem to yield beta rhythms, while combinations of delta and theta waves seem to come with sleep.

Some researchers believe that by learning to duplicate these wave patterns ordinary people might become wiser and more original, relax without tranquilizers, control fears, and sleep more restfully. Some success has been achieved in teaching people to produce alpha waves by use of bio-feedback principles. The method is very similar to that employed by those working on bio-feedback control of heart rate. In this case, however, the subject is informed that he is producing alpha waves by means of a warbling sound produced in the headphones he wears during training sessions.

The field of bio-feedback training is still very young. New research efforts in it are beginning at universities and laboratories all over the world. Most of the people who are now working in this area are reluctant to speculate about what the future holds, but they are confident that the new knowledge about the hidden feedback loops in the body will stimulate much more research into the astonishing ability of human beings to learn.

8 Feedback and Stability in Social Systems

MAN'S APPLICATION of the principle of feedback to machines has profoundly changed the world. It has given us "automatic factories" that run almost without human intervention. It is now a well-established fact in medical science that almost every major system in the human body is governed by a complicated feedback loop that causes the body to operate at an equilibrium point, or in "homeostasis." But, without doubt, one of the most fascinating areas to look for the feedback principle at work is in social systems. Whenever men or animals live together they form a social system that in some respects acts like any dynamic machine. That is, feedback loops are created that tend, in most cases, to drive the system toward an equilibrium point where the various forces are in balance.

Animal Populations

Evidence of feedback can be found by studying the fluctuations of animal populations in a given territory. Sometimes these interactions are very complicated. Charles Darwin, a great English naturalist of the nineteenth century, used a feedback mechanism to explain why there are more bumblebees near towns. His explanation was that near towns there are more cats which means there are fewer field mice and field mice are the chief attackers of bees' nests. Thus near towns, bees can enjoy more protection and their numbers can increase.

Sometimes when one animal species depends on another, it can produce an oscillation in the population of those two species. To illustrate this point, but leaving out the complications that are present in any real situation, consider a territory inhabited by squirrels and foxes where the squirrels are the chief food of the foxes. When the squirrel population is large, the fox population will increase. But as the number of foxes grows, the squirrel population

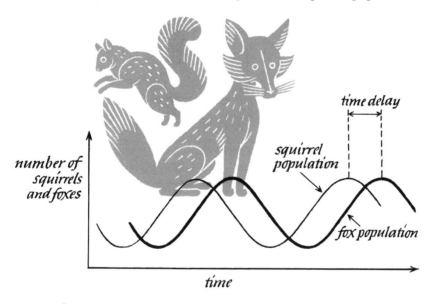

FIGURE 18

The population of squirrels in a region is related to the population of foxes in an oscillatory way. As the squirrel population increases, the food supply for foxes increases, thus causing the fox population to expand. But soon the increased fox population takes its toll on the squirrel population, causing the number of squirrels to diminish. This produces a scarcity of fox food, causing the fox population to follow the decline in the squirrel population. The interval of time between when the squirrel population becomes a maximum (or a minimum) and when the fox population achieves its maximum value (or minimum value) is called the *time delay*. Mechanical systems can also have time delays built into them.

will begin to fall, because more squirrels are caught. Then as the squirrels diminish, the foxes go hungry and their number declines. This results in a self-maintaining oscillation, or fluctuation, sustained by negative feedback with a time delay. Usually the time delay is such that the oscillation of the population is stable. This is not the complete story of the phenomenon known as the "fur cycle" which occurs in Canada, but it does illustrate an important point in the mechanism that causes it.

But what about when man interacts with man? There is no bigger area of interaction between men than in the marketplace, and so that is the next area that will be examined for evidence of feedback. A free market is commonly defined as one made up of both buyers and sellers who are not being forced to make purchases or sell goods. When you go into the supermarket to buy soda pop, you and the supermarket are assumed to make up a free market. You don't have to buy the package of soda pop and the supermarket doesn't have to sell it to you. In such free-market situations the law of supply and demand is presumed to hold.

The Law of Supply and Demand

An expression frequently heard in discussions on economic subjects is that "supply always adjusts to meet demand." If sales of new automobiles in the country are strong, the auto producers will increase production to satisfy the higher rate of sales. If sales are slow, production is cut so that the inventory of new cars does not get too large.

The "law of supply and demand" clearly involves the concept of operating around some equilibrium or static state. In economic systems, as in all other situations involving people, the location of this equilibrium point is not a simple thing and many factors influence where the static state will actually be. But first, what is the one factor that influences supply and demand the most? If you think about it awhile you will probably come to the conclusion that it is *price* that affects supply and demand more than anything else.

The law of supply and demand can be stated in the following way: the *demand* for an item decreases as its price increases. The *supply* of an item usually increases as its price increases. Furthermore, the law implies that a stable price is achieved if, and only if, the supply is equal to the demand. An example of how price and quantity are related for a buyer (or demander) is shown in Figure 19. Suppose that this drawing shows how the price of a bicycle would change depending on how many a buyer would want to purchase from a manufacturer. The more bicycles the buyer wants the less he will have to pay for each one. Now, think of the manufacturer (or supplier) of bicycles. The more money he can sell his bicycles for the more anxious he is to make bicycles. So he relates his production of bicycles to their selling price as shown in Figure 20.

Now when the buyer and the seller get together, their desires are satisfied at some quantity and price as shown in Figure 21. At this stable or equilibrium point the manufacturer (say the XYZ Bicycle Company) is happy to make two hundred and twenty bicycles for the purchaser at a price of twenty-three dollars per bicycle. Likewise the buyer of bicycles (say the ABC Department Store) is happy to buy two hundred and twenty bicycles at twenty-three dollars per bicycle.

The feedback principle can be used to describe this system by identifying three basic elements in the system:

> The supplier or manufacturer of bicycles
>
> The demander or buyer of bicycles
>
> The pricer or market place which acts as the comparer and reducer to set the price at which bicycles will be bought and sold.

The pricer in this case might be a giant bicycle sales show held in a large coliseum twice a year. At this show buyers of bicycles such as the ABC Department Store attend to shop for bicycles for Christmastime and springtime sales. Manufacturers of bicycles such as the XYZ Bicycle Company also attend to offer their products to the buyers of bicycles. Unlike earlier examples the pricer, which acts as a comparer and reducer, cannot be represented as a piece of physical equipment like a thermostat or furnace. The market place helps to set the price at which bicycles will change hands because it brings together in one place lots of manufacturers like XYZ who wish to sell bicycles with lots of purchasers like ABC who wish to buy bicycles. It does not take a central market place like a bicycle show very long to set a price which will cause the XYZ Bicycle Company to supply a certain number of bicycles (which can be found from Figure 20) and cause the ABC Department Store to buy a certain number of bicycles (which can be found on Figure 19).

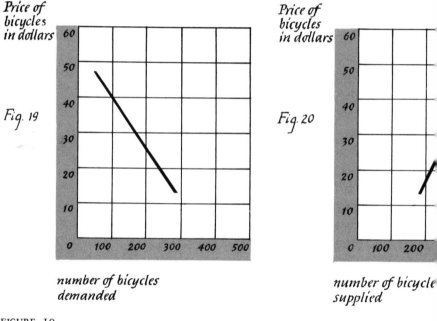

FIGURE 19
The relationship between price and quantity for a buyer who wishes to purchase bicycles. The more bicycles the buyer wants, the less he is likely to pay for each one.

FIGURE 20
The relationship between price and quantity for a manufacturer of bicycles. The more money he can get for each bicycle, the more bicycles he will manufacture.

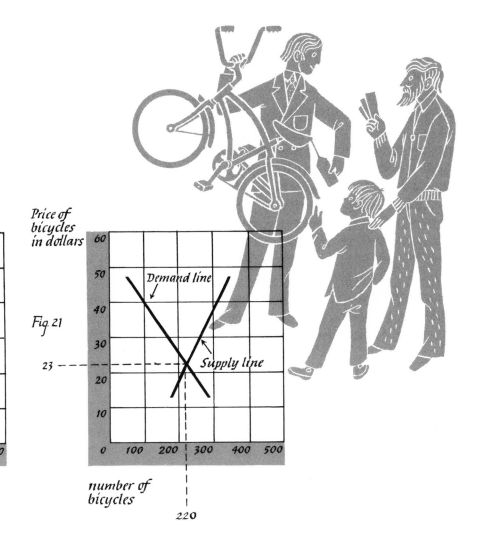

FIGURE 21
In a market situation the buyer and seller meet, causing the manufacturer's supply line to cross the buyer's demand line. Where these two lines cross is the equilibrium price that satisfies the wishes of both the buyer and seller.

If the number of bicycles the XYZ Bicycle Company and other manufacturers are willing to make does not equal the number of bicycles the ABC Department Store and other purchasers want to buy, the pricer makes a change in the market price in a direction which will make the supply eventually equal to the demand. Because of feedback the system will eventually come to an equilibrium price because it is that amount which brings demand for bicycles in line with the supply of bicycles. Thus both the buyer and seller must be considered in the feedback loop since they both determine what control action will be taken (if any) by the pricer. Figure 22 illustrates the feedback loop for this market system.

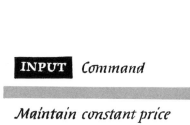

INPUT *Command*

Maintain constant price

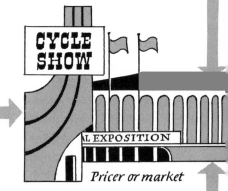

Pricer or market

This example does, of course, oversimplify how the market place operates because no market is nearly so predictable as the one described here. But even though it does not work exactly every time, it is useful to see feedback in an area so far removed from the areas discussed earlier such as the ancient water clock, flyball governor, and the automatic airplane landing systems now being developed.

But even when man does not necessarily want to interact with his fellow man, he cannot keep feedback out. Certainly when an infectious disease starts in a community no person wants to contribute to its spread. But exactly what does cause epidemics to start to spread? And to stop?

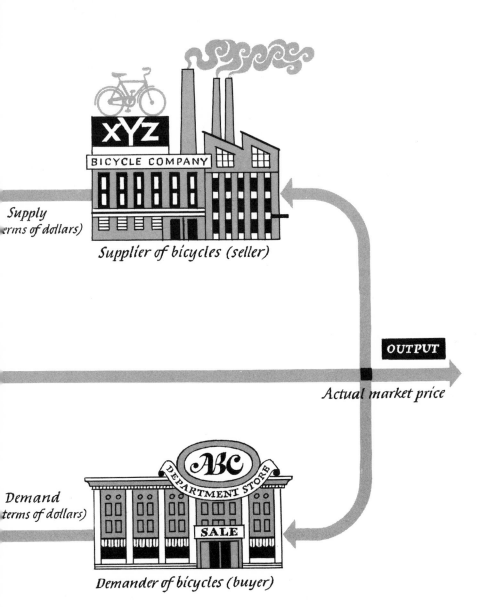

Supply
(terms of dollars)

Supplier of bicycles (seller)

OUTPUT

Actual market price

Demand
(terms of dollars)

Demander of bicycles (buyer)

FIGURE 22
Feedback loop for an economic system. The market (comparer and reducer) makes the necessary adjustments in price to match supply with demand and achieve a goal of constant prices.

9 The Role of Feedback in Epidemics

AN EPIDEMIC is the spread of a disease through a population at a very rapid rate. Whether or not a disease carried by a few people will grow into an epidemic is really a question of the stability of a feedback system.

Several years ago a person suffering from smallpox was accidentally permitted to enter the United States. After learning of this fact, public health officials began an immediate search to locate all persons who had been in contact with this individual since his arrival. They recognized how important it is to isolate or quarantine infected individuals to prevent an unstable situation from developing. The single infected individual acts as an initial disturbance to the entire community, which is originally in equilibrium.

Just as the development of the steel beam and the elevator produced a dramatic change in the construction of office space, so too is an epidemic the outgrowth of an abrupt change from one state to another; hence it is also an unstable phenomenon. The disturbance in an epidemic can be a single infected person in a population, a polluted water supply, or a shipment of bad food carrying the germ or virus that will cause the disease.

For a number of years public health officials have been working on developing a mathematical model or picture of how an epidemic actually works. Think of a community in which many people are immune to smallpox because they have been vaccinated. What happens if a single infected person comes into this community? He may infect some persons. But if these come into contact only with immune persons no epidemic will develop. If, on the other hand, each of these infected persons is able to infect three or four others, then an epidemic is on and it will grow very fast.

With many diseases people who recover become immune to that disease—they are no longer susceptible to catching it. If they recover quickly enough, they may not infect enough additional people to expand the epidemic. How important are these various factors relative to one another in the growth of an epidemic? One of the questions public health officials must try to answer then is, should they concentrate their resources on quarantine or on vaccination? That is one of the purposes of trying to build a model: to simulate an epidemic. Because epidemics are not predictable and are subject to many unpredictable disturbances, the models that are used never give absolutely correct answers. But such models are useful in that they can provide insights to how an epidemic becomes unstable and what policies may best bring it under control.

Here is a very simple example of an epidemic model. Assume that in a total population of N persons, everyone either is already infected with a certain disease or is susceptible to catching it. This model does not include the possibility of individual recovery and subsequent immunity, death, or isolation. The model has been purposely over-simplified so that the feedback and stability ideas become clearly visible. Such over-simplification usually does not occur in real epidemics except in certain mild chest infections with a long period of infection.

The number of persons in the population N who are already infected will be called I. The number of persons who have not caught the disease but are susceptible will be called S. The model that is being sought is one that will predict how many new persons will become infected each day. The rate at which new persons will become infected will be called *R*. It is expected that this number will depend on how many persons are already infected, for the greater this number the more likely that they will transfer the infection to others. But it is also reasonable to suppose that the number of newly infected persons on any day will depend on the number of persons who are susceptible to becoming infected. Once again, the greater the number of susceptible persons, the greater the number of persons who will become newly infected.

Now the model builder must make an assumption on how an epidemic spreads. He must make a guess on how the rate of infection (R) depends on the number of people who are already infected (I) and the number of people who are susceptible (S). One of the most obvious models that could be imagined is to suppose that the number of newly infected persons depends on the product of the number of people already infected and the number susceptible; that is, R is proportional to the product of I and S. The model builder might assume, for example, that the newly infected number is one percent of the product of the susceptible and infected number:

$$R = 0.01 \times S \times I$$

Now, in this model, if the number susceptible is zero there will be

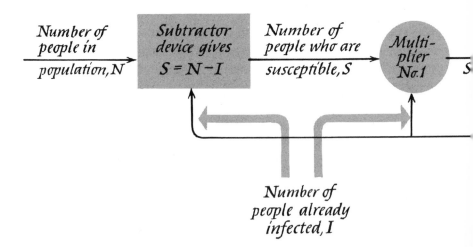

Number of people in population, N → *Subtractor device gives* $S = N - I$ → *Number of people who are susceptible, S* → *Multiplier No.1* → S

Number of people already infected, I

no newly infected persons and if the number of persons infected is zero that also would prevent any increase in the number infected. This model thus seems to be consistent with what we know about epidemics and is indeed one of the models used to study and predict the nature of epidemics.

Now comes the task of drawing a feedback loop of an epidemic situation. Since there are no physical quantities that will actually be controlled, it is necessary to think about what is involved a little differently. The number that is being sought is the actual number of persons infected, which is assumed to be the output. The input is the total number of persons in the population. The output does, of course, influence the number of new persons infected, so that will be the quantity that is fed back.

Figure 23 represents the feedback control loop of an epidemic. Imagine that the elements that influence the spread of an epidemic can be represented in boxes as shown. Diagrams similar to this one are used to help researchers wire up special types of calculators called analog computers. They are called analog computers because an electrical voltage is made to be equivalent or analogous to the quantities being studied. The desired answers can then be read on voltmeters and turned back into the quantities sought. The

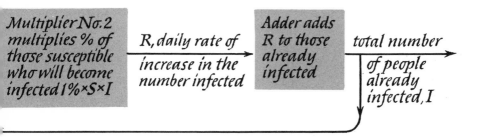

FIGURE 23

A feedback loop of an epidemic. This over-simplified diagram shows the important factors that influence the spread of an epidemic. This loop says that one can find out how many people will have been infected so many days after an epidemic begins, given a certain size population and a certain daily rate (in this case one percent) at which people will become infected.

boxes in the diagram correspond to parts that can actually be wired into the computer.

The first box is called a subtractor. It takes the number of persons already infected (I) and subtracts it from the total number of persons in the population (N) to get the number who are actually susceptible (S). The next box is the first multiplier, which multiplies the number susceptible times the number infected—quantities which have been fed back from the output. That quantity ($S \times I$) is then multiplied times the one percent in the next box to get the daily rate of increase in the number of infected cases. That information is fed into an adder which has kept track of all the people previously infected. When the newly infected number R is added to what is already stored in the adder (the number of people already infected), the output represents the total number of infected cases after any period of time—that is, the quantity I.

If actual numbers are used in this example, it is possible to see how an infection spreads. Figure 24 plots the number infected on the vertical axis and the number of days after the infection started on the horizontal axis for a population of a hundred people. After ten days everyone has been infected.

A more realistic model of an epidemic is one which takes into account the fact that people also recover and become immune, or die, or can be isolated from the rest of the community or leave the area in which the epidemic is spreading. For example, if one supposes that ten percent of the population will be withdrawn from the ranks of the infected because of recovery, death, or the like every day, one can observe how an epidemic can peak. The feedback diagram for this becomes somewhat complicated, but we can still see what happens to an initial population of a hundred as shown in Figure 25. Note that at no time does more than about one-half of the population become infected. This happens because the number of susceptible people falls very quickly. Thus it is clear that people who gain immunity from having had the disease exercise a big influence on the number of people who will actually be infected. If a feedback diagram were drawn for this model—similar to the diagram drawn for the economic example presented earlier in that it has a double feedback loop—one of the loops gives the number recovered each day while the other gives the total number infected. Feedback thus makes itself felt in all epidemics—from a mild winter flu infection in the United States to a major cholera epidemic in Africa.

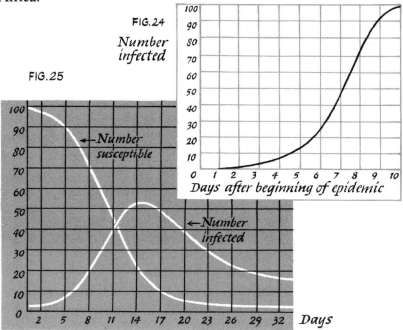

FIGURE 24
The results of the calculations made by using the model illustrated in Figure 23. This graph shows the total number of people who have become infected on any given day after the epidemic begins. These results show that on the tenth day everyone will have become infected.

FIGURE 25
A more realistic model of an epidemic is one that takes into account the fact that people may recover and become immune, or die, or can be isolated from the epidemic. Note that with this change in the model not everyone becomes infected and the number who are susceptible to the disease by the twentieth day becomes very small. (Calculations for both models were taken from *The Man-Made World*, Part Three, New York: McGraw-Hill Book Company, 1968.)

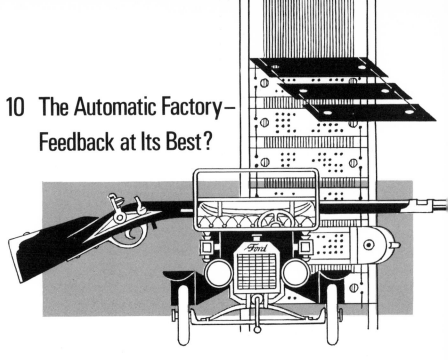

10 The Automatic Factory—
Feedback at Its Best?

As the steam engine and the flyball governor quickly spread from industry to industry—first in England, then the rest of Europe, and then throughout the world—old ways were quickly thrown aside. The Industrial Revolution was on. The widespread availability of power, not just on streams and rivers where water mills could be built but anywhere, had a profound effect on almost every aspect of life. Men and women were freed from their narrow village life and moved in ever increasing numbers to the rapidly changing cities, where supposedly they would find positions and wealth they could never have dreamed of before. Unfortunately the transition from peasant village to big city living produced almost as many problems as it created. People left their poor villages to live in crowded, bleak slums of the industrializing cities. There, instead of being subjected to the whims of the weather's effect on their crops, they became subject to the whims of the economic market place for the goods produced in the grimy factories that they toiled in twelve to fourteen hours a day.

The Industrial Revolution widened the gap between rich and poor, between novelty and custom and between town and country; it took more than a hundred years to narrow that gap appreciably. Not all of the inequities created by the Industrial Revolution have been eliminated, but on the other hand this sudden change in man's condition is not very old. For ten thousand years man had only his hands and occasionally animals to do his work. And then less than three hundred years ago, James Watt, Matthew Boulton, and others provided him with an inanimate source of almost unlimited power. Some thoughtful persons believe that we are already in a second industrial revolution in which man does not use machines but actually has machines use machines. The process of removing man from the feedback loop and replacing him with a mechanism is sometimes called *automation*. The ability to build automatic machines has been both hailed as a great achievement of man's abilities and attacked as a threat to society as it is known today.

Automation is viewed as a wonderful example of man's ingenuity because it will ultimately free him from doing all, or almost all, of the dull, dangerous, or boring jobs that must now be done in an industrialized economy. It is viewed as a threat by some who claim that if you keep eliminating jobs, sooner or later there will be nothing left to do. These pessimists believe that automation could take away most people's ways of earning a living. Even if man did not lose all of his employment he might wind up working only a few hours a week. They feel that many people might not be able to live a very satisfying life if they were employed only ten or twenty hours a week. The problems raised by automation are certainly not simple ones and they will continue to offer challenges to both engineers and social philosophers all over the world for a long time to come.

Why can man now build machines, factories, and offices that practically run themselves? The answer is found again in the universal phenomenon of feedback. It is only because machines can be made to check their work and compare it with what is desired that such systems can be built. But before examining the role of feedback in an automatic factory it is instructive to consider briefly the history of automation.

The Beginnings of Automation

Before the Industrial Revolution was well underway, Oliver Evans in 1784 built an automatic factory just outside of Philadelphia. It processed flour on a continuous basis—not in batches as other millers did—making use of powered conveyors to move the flour. No human labor was required from the time the wheat was received at the door of the mill until it was ready to be delivered to the customer as finished flour. In Paris in 1801, Joseph Marie Jacquard demonstrated an automatic loom controlled by—would you believe?—punched paper cards, similar in many ways to the punched cards used in modern office equipment. Thus the first tentative steps toward the automatic factory had begun with no one really realizing what the consequences of these steps would be.

If one wants to build a factory or a plant to run automatically it is almost a necessity that the plant produce many identical or nearly identical parts. People who own and operate factories want to produce thousands or millions of the same items so that the plant can operate without being stopped frequently. When a plant is not operating, most of the costs of owning the plant continue while no saleable products are being produced. The key to successful automation then is mass production.

Almost everyone has heard of Eli Whitney because of his invention of the cotton gin. It was not, however, the major achievement of his life. Whitney demonstrated the practicability of interchangeable parts in manufacturing which is another necessary key to automated production.

In 1789, shortly after Watt started using the flyball governor, Whitney offered to build for the United States Government ten thousand muskets. Up until that time guns had always been made one at a time by a gunsmith. He would make all the parts and then fit them together. There would be a great similarity among the guns made in the same shop, but the parts of one gun would not fit another, even when the two guns had been made by the same gunsmith.

Whitney looked at this problem and then reasoned that it would be more efficient to have one man, or group of men, make a particular part, using power tools as much as possible, and then assemble the guns from the baskets of stocks, barrels, triggers, and so on. He was brought to this method not just because he realized it would be faster, but because of the shortage of skilled mechanics available to build guns. This is one of the same arguments used for automation today. Whitney expected to use mainly not very skilled workers in his plant.

Whitney was not the first person to think up the idea of making up batches of parts and then assembling the final product from the various batches. But he so advanced the idea of *interchangeability* that his work overshadowed everything that had been done in this area. He worked out specialized devices for the job, called *jigs*. A jig is a shape or pattern of wood or metal through or along which a tool moves to produce copies of the shape of the jig. A coin can be used as a very simple jig for drawing a circle. A metal sheet with holes punched in it can be used as a jig for drilling holes in a piece of work much faster than if the holes were laid out on each separate piece of work. Whitney also built, and in some cases invented, special machine tools for the kind of mass production he had in mind.

The guns Whitney made in his shop were not as good as the best guns which could be made by a gunsmith but they were adequate and, more importantly, they could be turned out fast and inexpensively. Though Whitney had taken a giant step forward toward true mass production he apparently could not think of a better method of assembling guns than shuffling his interchangeable parts around the factory for the various operations required. This practice was to continue for another one hundred years. But late in the 1880's an efficiency-minded foreman in a steel plant, Frederick Winslow Taylor, began a systematic study of what goes on in industrial plants. He used time-and-motion studies to see how workers actually spent their time, and systematized cost accounting to find out what a product really costs. His work was finally capped by Henry Ford's power driven moving-chassis assembly line which eventually led to the automated production line.

The Ford assembly line was first installed at the Highland Park plant in Detroit in 1914. In building an automobile it is necessary to assemble numerous parts and subassemblies into a highly complicated machine. Ford used his and others' ingenuity to devise a method to combine pre-existing elements into a main assembly line fed by smaller lines, all tightly synchronized and coordinated. Their efforts produced an automobile which almost any American could own or hope to own and which could be mass-produced with steadily increasing output and for a while steadily decreasing prices.

This assembly line produced fifteen million cars up to the time the Model T Ford was discontinued in 1927. A kind of automation was in use but it was not automation as it is understood today. In this kind of automation men were used as machines. Some tentative efforts were being made to replace the automatons of flesh and blood with automatons of metal. It was at the Morris Motors plant at Coventry, England in 1924, that the first transfer machine was installed. It consisted of a series of automatic machines in a line which could transfer the part being worked on from one machine to the next. As the workpiece—castings which were being turned into engine cylinder blocks—arrived at each station, a certain job would be done on the casting; the casting would then be transferred to the next station where another job would be done on it. The machine worked but it was ahead of its time. It failed to save any money and so was broken up into separate units. Thus automation as it is known today was tried—and failed—even before the Model T passed from the scene. About twenty-five years passed before automation was again thought about in a serious way by people who worked in production and by those who worried about the effects of machines on society.

The Lifeblood of Automation

Stop and think a moment about all the earlier examples of feedback that have been presented so far—the automatic pilot, the thermostat, the control of the human body temperature, the price of land for skyscrapers, and the teaching machine. In each case what is actually fed back around the loop? In each case the answer is *information,* because that is the only thing any system can use to correct itself. An automatic landing system must have information about where the airplane is with respect to the runway, just as a thermostat must know what the temperature of the house actually is if it is to keep a desired temperature in the house.

Though man has been handling information for thousands of years—mostly in spoken form but more recently in written form—he was really limited by the size of his brain to thinking about one thing at a time. Though the human brain is probably the most re-markable decision-making device known, it still has trouble in handling more than three or four inputs at a time and it quickly becomes bored when asked to do the same tasks over and over. It was when man learned to build machines that could handle hundreds or even thousands of streams of input data, and then remember large amounts of this data that he was able to use feedback to automate factories in a very complete way. That is, it was his ability to construct very large and very reliable computing machines that has brought automation to so many industries. For example, what happens when a computer is designed to run a single complex machine?

Suppose the machine is a milling machine. Such a machine is designed to form metal into complex shapes. Normally a milling machine operator is given one or more drawings which might show the part to be made as seen from several different directions; he might also receive an operations-breakdown sheet telling him in which order to make the cuts on the part to be shaped, as well as some oral instructions from the shop foreman or the part-designer himself. It is precisely because communication between men, and between men and machines is so slow and complex that it pays to bring in data processing machines.

When one machine communicates with another, the instructions are necessarily in physical form—usually as pulses of electricity. The control of the milling machine actually starts back in the engineer's office where all the information about the size and shape of the part may be read off drawings. All of this information, in the form of numbers, is recorded on paper tapes in a tape-punching business machine. A simple milling machine might use this paper tape directly, but in the more complex example the paper tape is run through a computer-type machine which translates the numerical information into equivalent electrical signals. These signals are then recorded on a single magnetic tape much like that used in a home tape recorder. However, this recorder has fourteen channels and can record as many as fourteen tracks of information on the tape. The tape now contains all the numerical information that was originally on the drawing. See Figure 26.

This tape can now be sent to any factory equipped with the appropriate machines for handling information from magnetic tapes. It has been suggested that at some time in the future manufacturers might not try to ship spare parts all over the world (especially big ones), but might just ship magnetic tapes to factories near the place where the parts are needed. In fact it is not even necessary to ship the tapes since the electrical signals on the tapes can be transmitted over an ordinary telephone line to a waiting tape recorder.

At the factory, the tape with the information on it is now inserted into a playback device which will reproduce the electrical signals which were originally placed on the tape. These signals from the fourteen tracks, or as many of them as needed, are fed into a group of amplifiers and power controllers which guide the movements and cutting action of the cutting tools.

But that is not all. Suppose that the cutting tool is not as sharp today as it was yesterday when the tape was used to operate the milling machine. Built into the machine is a measuring device that checks the dimensions after each cut is made. If not enough metal has been removed in the shaping process, it will order additional cuts and the comparer will specify how much more is to be removed to make the part come out the correct size.

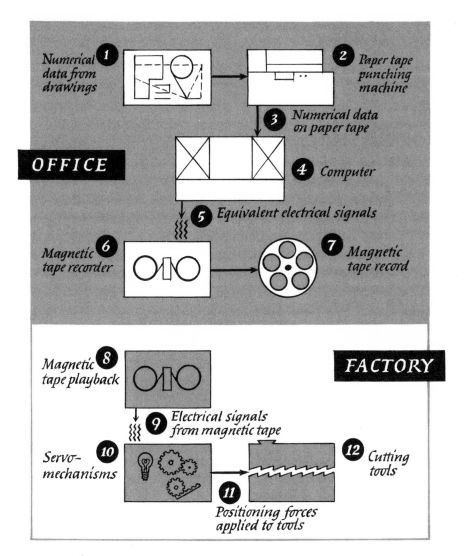

1 Numerical data from drawings
2 Paper tape punching machine
3 Numerical data on paper tape
OFFICE
4 Computer
5 Equivalent electrical signals
6 Magnetic tape recorder
7 Magnetic tape record
8 Magnetic tape playback
FACTORY
9 Electrical signals from magnetic tape
10 Servo-mechanisms
11 Positioning forces applied to tools
12 Cutting tools

FIGURE 26

How an automatic machine tool works is illustrated on this diagram. Numerical data taken from an engineering drawing is translated by a computer into electrical signals and then stored on magnetic tape. The magnetic tape is then played on a special tape recorder which translates the information on the tape back into electrical signals which are fed into a servo-mechanism-driven machine tool. This tool now positions the cutting edge of the tool to make the correct cuts on the workpiece. The unit will also have built into it a feedback loop to correct the position of the cutting tool if it is failing to shape the piece to the accuracy required by the original drawing.

FIGURE 27

Photograph of a portion of the automatic engine manufacturing line at Ford Motor Company's Cleveland plant. Engine cylinder blocks are moved by a transfer machine (at the left) to a broadside shuttle (at the center) which then feeds blocks to the Sundstrand mill where further machining operations will be carried out on the blocks. If one mill has sufficient blocks to work on, the transfer machine will automatically by-pass it and move the block to a second mill. Photograph courtesy of the Ford Motor Company.

Without this last step the milling machine might be able to produce parts all day but they might be all the wrong size and completely worthless. It is the feedback portion of the machine that will ensure that the parts come out as originally specified on the drawing. As complicated and expensive as this example might sound, most manufacturers believe that they must go to automation of their routine operations or fail to compete in the market place.

The real potential of automation becomes clear, however, when one ties together a number of automated machines to do a series of different operations on a workpiece. This has come to be known as "Detroit automation." Specially designed metalworking machines, for example, are connected by a conveyor system that accurately positions the work as it is carried from station to station. At the Ford engine plant in Cleveland, rough castings of engine blocks are moved in and out of twenty or so different machines as more than five hundred separate operations of milling, boring, honing, drilling, and tapping are carried out (Figure 27). In this way only a dozen or so men monitor the production of hundreds of engines per hour.

Plants such as these are still primitive, however, when compared to a modern chemical plant or oil refinery. These two sectors of industry have spent more effort on developing processes than the firms in any other industry. In fact, they had no choice. In these two industries one does not have a product until one comes up with a process to make it. Both industries are mostly of the continuous flow type rather than a batch process industry. That is, there is a continuous stream of the final product always being delivered by the plant. These industries had to learn to handle huge volumes of gases and liquids at high velocities and temperatures. The "tools" used in the refinery and chemical plant are fractionating towers, which accurately separate desired chemical compounds from those which are not desired, and catalytic reactors that convert compounds of low value into compounds of higher value. Because they did not have a hundred or more years of tradition to fall back on as the machine tool industry did, these industries found new and ingenious ways to

FIGURE 28
The purification columns of a modern chemical plant—this one in Chicago makes 130 million pounds per year of phthalic anhydride. Such plants because of their high degree of automation generally employ very few people. Photograph courtesy of Koppers Company, Inc.

add to the capacity, flexibility, and precision of their tools (Figure 28).

It is hard to overestimate the need for automation in these two industries. There is a story told about an Asian country that hired a U.S. contracting firm to design a modern oil refinery. The firm submitted a design for the refinery and it included the usual array of control equipment. The officials of the country examined the plans and then asked the designers to eliminate all the automatic controls from the plant as the country had a considerable surplus of manpower. The officials explained that they could provide thousands of men, if needed, to record instrument readings and to control processes. They were also willing to accept a lower plant efficiency and poorer quality products to create employment. The designers listened sympathetically to this request but explained that the inclusion of automatic controls was not based simply on operating costs or efficiency: without suitable control instruments (and, of course, lots of feedback loops) a modern refinery simply could not operate at all.

And What of Tomorrow?

Norbert Wiener, one of the great thinkers about the consequences of feedback, as mentioned earlier defined it as "the property of being able to adjust future conduct by past performance." But less than thirty years ago there were no truly automatic factories or chemical plants, so how can the past help us to learn to cope with a world which is becoming more automated? It will not be easy and so far no one has come up with any simple answers to the problems that automation has created regarding man's work and his leisure time. But as with almost all technological processes once you start down the road there is no turning back. Man must learn to live, labor, and play in a world of automatic factories, plants, and offices.

There is a fable told about a Hindu wiseman which might help us to examine thoughtfully the problems that have been created by feedback and the automatic plant. The wiseman had been granted by Heaven the ability to create clay men. When he took earth and water and fashioned little men, they lived and served him. But they grew very quickly, and when they were as large as himself, the wiseman would write on their foreheads the word DEAD, and they would become dust. One day he forgot to write the word on the forehead of one of his full-grown servants, and when he realized his mistake the servant was too tall—his hand could no longer reach the slave's forehead. This time it was the clay man's turn. He promptly killed the wiseman.

Let us hope that this fable does not apply to us and that we will learn to handle feedback loops both wisely and well, and that these loops will lead us to a world of plenty, free forever from the mind-numbing tasks much of mankind must do everyday just to survive.

BOOKS FOR FURTHER READING

Annett, John, *Feedback and Human Behavior*. Baltimore: Penguin Books, 1969.

Automatic Control, by the Editors of Scientific American. New York: Simon and Schuster, 1955.

Calder, Nigel, *The Mind of Man*. New York: Viking Press, 1970.

Diebold, John, *Automation, the Advent of the Automatic Factory*. New York: Van Nostrand, 1952.

Doebelin, Ernest O., *Dynamic Analysis and Feedback Control*. New York: McGraw-Hill Book Company, 1962.

The Man-Made World, Part Three. New York: McGraw-Hill Book Company, 1968.

Mayr, Otto, *The Origins of Feedback Control*. Cambridge: M.I.T. Press, 1970.

Ruch, Floyd L., and Zimbardo, Philip G. *Psychology and Life,* 8th Edition. Glenview, Illinois: Scott, Foresman, 1971.

Index

ABOUT THE AUTHOR

Stanley W. Angrist's first two books for children, *How Our World Came To Be* and *Other Worlds, Other Beings,* explored the creation of the solar system and the possibilities of life elsewhere in the universe. As he began to delve into the area of feedback, he realized how pervasive a phenomenon it is, which led him to extensive research on another fascinating subject for young people (and older ones, too).

Dr. Angrist is Professor of Mechanical Engineering at Carnegie-Mellon University. He lives with his wife, Shirley, and his three sons, Joshua, Misha, and Ezra, in Pittsburgh, Pennsylvania.

ABOUT THE ILLUSTRATOR

Enrico Arno has had a distinguished career as an illustrator of children's books. He was born in Mannheim, Germany, and educated in Berlin. In 1940 he emigrated to Italy, where he worked for book publishers in Milan and later in Rome. Mr. Arno came to the United States in 1947. He lives with his wife in Sea Cliff, New York.